The Credibility Crisis in Science

The Credibility Crisis in Science

Tweakers, Fraudsters, and the Manipulation of Empirical Results

Thomas Plümper and Eric Neumayer

The MIT Press
Cambridge, Massachusetts
London, England

The MIT Press
Massachusetts Institute of Technology
77 Massachusetts Avenue
Cambridge, MA 02139
mitpress.mit.edu

The MIT Press would like to thank the anonymous peer reviewers who provided comments on drafts of this book. The generous work of academic experts is essential for establishing the authority and quality of our publications. We acknowledge with gratitude the contributions of these otherwise uncredited readers.

This book was set in Stone Serif and Stone Sans by Westchester Publishing Services. Printed and bound in the United States of America.

Library of Congress Cataloging-in-Publication Data is available.

ISBN: 978-0-262-05127-9

10 9 8 7 6 5 4 3 2 1

EU Authorised Representative: Easy Access System Europe, Mustamäe tee 50, 10621 Tallinn, Estonia | Email: gpsr.requests@easproject.com

It is easy to lie with statistics, it is easier to lie without them.

—Frederick Mosteller

Contents

Preface

Science has an extraordinary record of achievement. From landing humans on the moon to dramatically increasing life expectancy worldwide, its accomplishments have reshaped our understanding of the world and our place within it. These successes stem from science's remarkable ability to uncover causal mechanisms and solve complex problems.

Yet, despite its power, science is fragile. It requires protection not only from external threats, such as deniers and pseudoscientists who misuse it for their own agendas, but also from within. Scientists sometimes manipulate research outcomes to serve personal interests.

This book defends the integrity of scientific inquiry against such manipulation, focusing on three main strategies: data fabrication, where data are invented; data manipulation, where raw data are altered, typically in undisclosed ways; and "tweaks," where researchers adjust research designs or model specifications. These tactics share a common goal: to produce results that align with a predetermined narrative.

Whether for career advancement, personal gain, ideological bias confirmation, or media attention, these manipulations undermine the scientific method. Intentional distortions of empirical results, whether through fabrication, manipulation, or tweaking, should be treated with equal seriousness, regardless of the method used.

This book addresses data fabrication and manipulation but primarily focuses on tweaking, the often overlooked and, as we argue, more common strategy within the darker side of science: the deliberate manipulation of results through selective research design and model specification choices. The object of manipulation is typically the statistical measure considered most important in a study. Researchers may test various model specifications and selectively report only those that yield "publishable" results—those that meet a threshold for significance, such as a p-value below 0.05, a substantial effect size, or an impressive explanatory power.

A book typically makes more than one argument, and ours is no exception. Chapter 1 argues that the intentional manipulation of model specifications and research designs—what we call "tweaking"—is fraudulent, just as data fabrication and manipulation are. The intent is the same; the results are the same; only the methods differ. While there is widespread agreement that data manipulation is unethical and data fabrication is even worse, there is less, if any, consensus on whether tweaking also constitutes fraud. Many scientists view tweaking as a gray area, somewhere between honest research and outright fraud. However, this view trivializes the issue. Like data fabrication and manipulation, tweaking invalidates empirical results, often though not necessarily leading to false inferences. Tweakers, like outright fraudsters, manipulate empirical results for personal gain, hoping to boost their publication records and further their careers.

In chapter 2, we explore the motivations behind manipulating empirical results. We identify two primary drivers: increasing the likelihood of publication and aligning results with the authors' theoretical or ideological beliefs. Getting published is not the end goal in itself, but rather a means to achieve other self-serving objectives such as peer recognition, promotions, job offers, salary increases, media attention, and, in some cases, even the Nobel Prize. High-profile scientists have come under scrutiny for dubious data and questionable results, demonstrating that manipulation can affect even the most celebrated figures in science.

Acknowledging that tweaking is fraudulent has significant implications. As chapter 3 shows, tweaking is typically observationally indistinguishable from honest ways of improving one's research design or model specification, making it difficult, if not impossible, to prove intentional manipulation in any particular case. However, replication exercises have shown that results from replication studies often deviate systematically from reported results in the original studies, particularly in terms of statistical significance and effect size. These systematic deviations suggest that some researchers deliberately select models based on statistical outcomes. While data fabrication and manipulation are sometimes detectable by techniques discussed in chapter 4, tweaking is subtle and typically undetectable, making it an attractive method for those seeking to avoid detection. The challenge of proving fraudulent intent remains substantial, and in many cases, it will be impossible to determine whether results were deliberately manipulated or reflected honest, though controversial, research design and model specification choices.

Part II explores how different research designs can be manipulated. Though tweaks vary across designs, it is always possible to distort empirical results: No matter the research design chosen, there are always researcher degrees of freedom that fraudsters can exploit.

Part III discusses strategies for deterring, preventing, and detecting fraud. We emphasize two approaches we consider most promising: rigorous data transparency policies and robustness testing. While neither approach will necessarily deter or prevent fraud or ensure detection when it is committed, radical transparency does help since it imposes specific requirements on publishing authors. If nothing else, it facilitates replication studies and robustness testing by others. Of the two, we prefer robustness testing over replication studies. Replication studies take the research design and model specification of studies as given and try to replicate results based on these choices. Robustness testing, by contrast, modifies these choices in plausible ways, exploring the sensitivity of results to these alternative choices. Fraudulent results tend to be non-robust. This is in their nature: If they were strongly robust to alternative research design or model specification choices, they would not have to be arrived at by fraudulent means in the first place. Rigorous robustness testing will thus diminish the incentives to fabricate, manipulate, or tweak results. Together, data transparency coupled with robustness testing will reduce the prevalence of fraud and help protect the integrity of scientific inquiry.

We make multiple contributions to various aspects of the debate on scientific fraud. In our view, the most important contributions are the following.

First, we counter the widespread notion that scientific fraud should be defined only as the fabrication or undisclosed manipulation of data. What researchers do with data when they design their research and specify their estimation models can be just as fraudulent. In our view, scientific fraud is better defined by the intent to manipulate empirical results: regardless of how the results are manipulated. In fact, we argue that while data fabrication and manipulation are relatively rare, even though their full extent is unknown, tweaking is likely to be much more common. The real threat to scientific knowledge therefore does not stem as much from data fabrication and manipulation as it does from tweaking.

Second, and perhaps surprisingly to some, our definition of scientific fraud does not require the manipulated results or inferences based on the manipulated results to be false. Fraud follows from the intention to gain an advantage by manipulating empirical results. It does not matter whether the results or inferences based on them happen to be false or not. Consider the analogy of a counterfeit Rolex: selling a fake Rolex as the real thing is fraud, regardless of whether the fake watch is as good as the original or not. Of course, the vast majority of manipulated scientific results will be false.

Third, we argue and demonstrate that no research design is immune to manipulation. While the majority of fraudulent experimental studies appear to rely on the brute force of either fabricating or manipulating data—the case of Francesca Gino, which we discuss in detail in chapter 4, comes to mind—a similar impact on results can be achieved by seemingly innocent variations to research design or model specification. This is best known and understood for observational data analyses but equally applies to all experimental designs and all causal models or in fact any other research design not explicitly addressed and explored by us in part II.

Fourth, we propose a data integrity scheme that allows readers to immediately assess the transparency and accessibility of the data used in an analysis. It warns readers to maintain a healthy skepticism if, for whatever reason, the transparency and accessibility of data are limited and suboptimal.

Fifth, we also make a point that should be obvious but is frequently neglected or ignored, namely, that researchers need to go the extra mile to demonstrate data integrity when they collect their own data. The Gino case illustrates how easy it is to manipulate empirical results by adding, subtracting, or manipulating information in self-collected data. We strongly recommend that universities and research institutes take measures and provide infrastructure to require researchers to keep a secure, safe, and unalterable copy of the raw data. Of course, we are not against data cleaning, transformations, and manipulations as long as these are justified and disclosed, but the raw data must remain "as collected."

Sixth, we stress that science would be better served by focusing more on the question of the robustness of empirical results to plausible changes in research design and model specification, rather than focusing almost exclusively on trying to find and prove dishonest manipulations, which will often be difficult if not impossible to do. Fraudulent research will tend to score low on robustness, but so does much honestly undertaken research.

Less robust findings should carry less persuasive power than more robust research. For robustness testing to exert its full potential, it is not enough, however, to require robustness tests but then leave the choice of robustness tests exclusively to the authors. This solution would simply raise the potential for fraudulent manipulation to the level of selectively choosing and designing robustness tests. We suggest that journal editors and reviewers specify the robustness tests they want to see before a manuscript is accepted and ultimately published.

* * *

We would like to acknowledge those who, like the Data Colada team, speak out against the manipulation of empirical results. We also thank the anonymous whistleblowers, the students and scholars who replicate and reanalyze published studies, the journalists reporting on scientific misconduct, and organizations like Retraction Watch that track instances of fraud. We hope our work exaggerates the scope of the problem, but we fear it does not. Finally, we express our gratitude to the MIT Press, especially to our editor, Catherine Woods, for her unwavering support, and our production editor, Judy Feldmann, as well as to our friends and colleagues who provided valuable feedback on earlier drafts. Special thanks to Ray Duch for his permission to use experimental data from various countries. And a very special thanks goes to the fabulous Lee McIntyre, whose dedication to the scientific attitude inspired us throughout this journey.

I Is Science Broken?

> Fasten your seat belts. It's gonna be a bumpy ride.
> —Zombieland Rule #4

Sitting at the Table of Gods

On March 29, 1999, the cover of *Time* magazine showed a picture of Albert Einstein. Einstein looked old and tired, his forehead covered in wrinkles, his hair long and gray, in a famous photograph taken by the portrait photographer Philippe Halsman back in 1947. During the photo session, Einstein expressed remorse for his inadvertent role in the Manhattan Project, the initiative that ultimately culminated in the devastating bombings of Hiroshima and Nagasaki. The picture of a sad and tired Einstein became Halsman's most iconic photo.[1]

Time magazine hardly ever places the picture of a celebrity from a past historical period on its cover. But in 1999 the editors had good reasons to ignore this rule: The magazine had designated Albert Einstein as the "person of the century," a distinction that placed him above notable figures like Mahatma Gandhi and Franklin D. Roosevelt, who were the runners-up. A great honor for Einstein and for the profession he represents: scientists. And Einstein was not the only scientist on *Time*'s list of the 100 most influential people of the twentieth century: This list featured nineteen scientists, which made them the third largest professional group, just a shade behind politicians and industrialists. The twentieth century was the century of man-made political disasters. But the twentieth century was also the century of science, and Einstein was its figurehead.

Those days seem to be over. And they may never come back.

In the twenty-first century, the role and relevance of scientists have changed. Science is no longer triumphant: It is in the midst of a severe crisis. Public trust in scientific results and findings has dwindled,[2] and science does not know how to regain credibility. This crisis of science has many facets. More than anything else, however, it is a credibility crisis. The public no longer believes that scientists merely make honest mistakes on the long and winding road to truth. Scientists are increasingly seen as partial, ideological agents, activists in an armchair, or, worse still, simply fraudsters who fabricate or manipulate data and tweak the specifications of their empirical models to get their desired results.

The credibility crisis of science is not about scientific progress invalidating previously held scientific beliefs. That the predictions of widely believed theories clash with new empirical results is normal and an inevitable part of scientific progress. Some, perhaps many, scientific theories and ideas have been shown to be false. Einstein cannot be right without Newton being, to some extent, wrong. And Einstein was not entirely right either. He was wrong about quantum mechanics, about spooky quantum entanglement, and he failed to accept that the universe is not constant in size even though his equations predicted an expanding universe.

The credibility crisis of science is not, as such, about scientific ideas, theories, or models that turn out to be false. Rather, the crisis has been caused by scientists who deliberately publish overconfident, misleading, and often simply false empirical results based on research designs or model specifications they have intentionally specified to give the desired results. We call this practice "tweaking." In extreme cases, published results rely on manipulated or outright fabricated data. Whether tweaked, manipulated, or fabricated, the results often cannot be replicated—not even if replication analysts use identical research designs.

By itself, failure to replicate does not necessarily indicate and certainly not prove scientific fraud. Empirical results can vary for many reasons. However, replication analyses usually show that replicated effect sizes are, on average, systematically smaller and, at the same time, often statistically insignificant. If 90 percent of replications (Open Science Collaboration 2015) deviate from the original article in one direction that is less favorable to what the authors wanted to demonstrate, then these deviations are not innocent random errors or acts of nature. If the deviations were random, they would cancel each other out and their mean would be close to zero.

Instead, these deviations indicate that many of the published results are likely to have been tweaked, manipulated, or fabricated.

Tweaking is potentially more damaging to science in the long run than data manipulation and fabrication. Any particular tweaked empirical result is likely to have a smaller effect on the fabric of science than the cases of data fabrication and manipulation, but the cumulative effect of tweaking can still be larger than the cumulative effect of data fabrication and manipulation because these strategies are rare, while tweaking is common.[3]

Since Retraction Watch began monitoring and reporting retractions,[4] the number of retracted articles per year has steadily increased. Some of this is due to "bulk retractions" of thousands of articles published in so-called paper mills, where authors pay for the publication of fake articles. We are not interested in these retracted paper mill publications. We focus here on variants of data fraud, a subset of retractions that have also been steadily increasing.

In recent months, we have seen the spectacular retractions of Francesca Gino of Harvard University and Marc Tessier-Lavigne of Stanford University. And these are just the most recent cases, the ones that stick in the public mind for a while before attention is drawn to other, more spectacular cases of scientific fraud.[5]

Scientists no longer sit at one table with the gods. They have become mortals. Science is in the midst of a credibility crisis. Is science broken?

As always, the correct answer is: It depends. Whether science is broken depends, first and foremost, on the meaning of "broken." If broken means that science is no longer what it used to be, it may well be broken. If broken means that science is no longer capable of providing benefits to mankind, then, no, science is not broken. But there is plenty of room between these two extremes.

Some say science is broken: "An astonishing number of scientific studies, especially in medicine and the social sciences, are wrong. Statistical naïveté, poor practice, and outright fraud have meant that scientific journals have filled up with false information."[6] "Scientific research is fundamentally flawed, rife with published findings that often can't be reproduced or replicated by other scientists, otherwise known as the replication and reproducibility crisis."[7]

Some claim that science is not broken: "Science isn't broken, nor is it untrustworthy. It's just more difficult than most of us realize."[8] "It seems to me that yes 'Science is not broken,' but that doesn't mean it doesn't need fixing."[9]

And some say that those who play down the challenges ahead try to protect their own interests: "Conservatives in science naturally prefer the 'self-correcting' frame, since it implies protection of the status quo, and greater deference to authority."[10]

Do scientists ignore the widespread manipulation of empirical results, because they have settled comfortably in their ivory tower?

If science is a window to seeing and understanding the world, then this window is clearly not shattered to pieces on the ground. But the window is also not as new, shiny, and bright as it once used to be. Over the past decades it has cracked and here and there the glass is foggy. Still, our conclusion is not that science is broken. It is not beyond repair. But to keep going on with business as usual is not an option either. While public trust in scientific results has eroded over time,[11] not enough has happened from within science to stem this tide.

Possible Responses to the Credibility Crisis

Scientists have been caught on the back foot. Perhaps they were surprised by the severity of the credibility crisis. And they have not responded well. Still today, ten years after the *Economist* turned public attention to "wrong science,"[12] the academic response to its own crisis is lame. To make matters worse, scientists belittle the crisis by bringing forward a long list of ad hoc arguments that all intend to let the problem appear smaller than it actually is. And that is not a good sign. We see six common defensive arguments, but we make no claim for completeness.

1st line of defense:

Only the black sheep of the scientific community manipulate empirical results. Black sheep exist everywhere, not just in science, and the detected cases are merely a few bad apples (Broad and Wade 1983). Thus, scientific fraud is "a relatively minor problem in terms of the overall level of science."[13]

2nd line of defense:

"Science is self-correcting, in the sense that falsehood injected into the body of scientific knowledge will eventually be discovered and rejected" (Goodstein 2010: 2). "What makes science so powerful is that it's self-correcting—sure, false findings get published, but eventually new studies come along to overturn them, and the truth is revealed."[14]

3rd line of defense:

Researchers do not mean to do wrong. They often cherry-pick empirical results without fraudulent intentions: "Scientists who fiddle [with empirical results], just about all of them do, . . . aren't usually committing fraud, nor are they intending to. They're just falling prey to natural human biases that lead them to tip the scales and set up studies to produce false-positive results."[15]

4th line of defense:

Science is a difficult craft and art, errors occur, and they are factored into the inferences scientists draw from their results. "Science deserves respect exactly because it is difficult—not because it gets everything correct on the first try. The uncertainty inherent in science doesn't mean that we can't use it to make important policies or decisions. It just means that researchers should remain cautious and adopt a mindset that's open to changing course if new data arises."[16]

5th line of defense:

Scientists have already begun to develop techniques that help to identify fraudulent empirical results: replication studies, forensic statistics, techniques that help identify implausible data structures, robustness tests, and so on. The rising number of retractions demonstrates that the policies to single out the black sheep are actually working. As McIntyre believes: "Scientists were paying close attention to the problem and trying to fix it. . . . The response from the scientific community has been pitch perfect: We've got this. We may not have it solved yet . . . but we are investigating the problem and trying to solve it" (McIntyre 2019: 97).

6th line of defense:

Regardless of the black sheep and rotten apples, science is still astonishingly successful. "The number of dishonest scientists cannot, of course, be known, but even if they were common enough to justify scary talk of 'tips of icebergs,' they have not been so numerous as to prevent science's having become the most successful enterprise (in terms of the fulfilment of declared ambitions) that human beings have ever engaged upon."[17]

These lines of defense carry more than a grain of truth. They are correct, to some extent. But it is also true that the manipulation of empirical results persists. There is no proof that the rise in retractions can be explained by higher detection rates alone, as the optimists claim. It may well be that the number of fraudulent articles rises along with retractions and perhaps faster than the number of published articles. We can only see the tip of the iceberg, and the true magnitude of the problem remains unknown.

The question thus arises as to the appropriate response of the scientific community. Downplaying the problem is not a viable strategy. While it is not the case that fraudulent empirical results obstruct scientific progress in the broader context, they do have the effect of eroding public confidence in science. Science is not broken beyond repair, but it is time to turn the tide.

In our view, one shift in perception and two shifts in strategy are needed to effectively address the credibility crisis in science.

First, scientists need to recognize that the credibility crisis in science is due not only to data manipulation or fabrication, but also to what is commonly known as p-hacking, the intentional manipulation of p-values, and what we call "tweaking." Tweaking is the deliberate choice of research design and model specification to produce specific empirical results in the interest of the researcher. Tweaking is a fraudulent technique that is difficult to detect. Consequently, although impossible to prove, we expect tweaking to be a much more common phenomenon than data fabrication and manipulation.

Second, counter-fraudulent strategies need to focus on the identification of dubious and fragile research results, which are sensitive to plausible changes of the research design or the model specification. Lack of robustness does not prove that empirical results have been manipulated, but it characterizes weak results and unreliable evidence. Elsewhere, we have suggested that journal editors and reviewers have to identify robustness tests before they accept manuscripts for publication, and we renew this proposal here (Neumayer and Plümper 2017). This is important because researchers can as carefully select robustness tests based on the results these produce as they select baseline model specifications based on results.

Third, counter-fraudulent strategies do not depend on what lawyers refer to as reversal of the burden of proof—a legal concept that shifts the burden of proof from the prosecution to the defense or from the plaintiff to the defendant. However, scientists need to be able to demonstrate that their data are genuine, that their data have not been fabricated or manipulated with the empirical results in mind. In addition, the best evidence that authors have not tweaked is to demonstrate that results are robust to plausible changes in the research design and the model specification. Such evidence will clearly be much more persuasive if authors do not themselves select the robustness tests employed.

1 Scientific Integrity and Its Discontents

> I wasn't able to get the results the way I wanted them. I couldn't resist the temptation to go a step further. I wanted it so badly.
>
> —Diederik Stapel

Scientific Fraud and Murder

Is there a difference between murder and scientific fraud? Of course there is: Murder kills, scientific fraud does not. At least not normally.

However . . .

Paul Kammerer, an Austrian biologist who worked on the heritability of acquired characteristics in the tradition of Lamarck,[1] produced spotted salamanders to provide evidence for the Lamarckian theory. When he was accused of fabricating his data, he attributed the manipulations to a research assistant. He committed suicide in 1926.

In the scandal surrounding the scientific misconduct of Paolo Macchiarini,[2] patients died, but whether they died because Macchiarini chose a treatment based on fraudulent scientific results or whether they would have died anyway remains an open question.

Andrew Wakefield, an anti-vaccination campaigner, published an influential article (Wakefield et al. 1998) claiming that measles, mumps, and rubella vaccinations caused autism. The article was fraudulent: Wakefield analyzed only twelve cases that appear to be carefully selected, and he did not reveal his conflict of interest. It had been retracted in 2010, but by then the article and the press conference that Wakefield gave about his paper had already inflicted substantial damage by encouraging parents not to vaccinate their children. Despite being retracted and despite compelling evidence disproving a causal link between vaccination and autism, the article continues to fuel anti-vaccination campaigns. In all likelihood, some people who would have been vaccinated in the absence of the fraudulent Wakefield article were harmed. And some children have died as a result.

None of these cases, however, constitutes murder. Not in a legal sense. The vast majority of fraudulent articles do not kill, and none of the three cases qualify as murder in a legal sense. These three cases are exceptions, even outliers.

There is, however, one case that links scientific fraud to murder: the tragic case of a fraudulent Bangladeshi academic who had a critic murdered. This was murder, unquestionably. The academic was sentenced to death and hanged seventeen years later, in 2023.[3]

Despite this one single case, scientific fraud and murder seem to be worlds apart. The vast majority of fraudulent articles do not kill, intentionally or not. From a different perspective, however, the two have at least some aspects in common.

First, it is possible to get away with murder, and this also holds for data fraud. In murder, the chances increase if the killer makes the cause of death seem to look natural, which rules out a few options such as guns, knives, bombs, and weapons of mass destruction. Clever murderers prefer nicely staged accidents—suspiciously many Russian oligarchs have recently fallen out of a window—or perhaps the elusive, untraceable poison. The same goes for scientific fraud: The more the resulting manuscripts look like normal scientific papers, and the more the evidence they present appears plausible, the better the chances of the fraudster of getting away with data fabrication, data manipulation, or the intentional massaging of empirical results. When it comes to scientific fraud, we do not know whether the cases we see are the tip of the iceberg or just the tip of the tip of the iceberg.

And second, murder and scientific fraud are both premeditated. Without intent, the killing of another human being is called involuntary manslaughter, and scientific fraud is called an honest mistake. Thus, intention is a necessary condition for both scientific fraud and murder.

Defining Scientific Fraud

This brings us to a more subtle distinction between scientific fraud and murder: Intentionally killing someone is murder, regardless of the choice of murder weapon. Stabbing a victim in the back with a knife is murder, and intentionally poisoning someone's food is also murder.

This does not currently seem to apply to scientific fraud. Many scientists and scientific institutions accept the fabrication and the intentional manipulation of data in order to produce a certain result as fraud, but the intentional tweaking of empirical results, the fine-tuning of research designs or model specifications undertaken in the interest of the tweaker, does not usually count as fraud.

Data fabrication is the invention of information, data, out of thin air. Scientists fabricate data because it requires less time and effort than honest data collection and it guarantees the empirical results the scientist wants. Data manipulation is the unjustified and typically undisclosed alteration of data. Researchers who manipulate data replace certain values in a dataset with others in order to change the empirical results. Tweaking is a deliberate change in research design or model specification to shift the empirical results in a certain direction.

Lee McIntyre defines scientific fraud as follows: "Fraud is the intentional fabrication and falsification of the scientific record." He distinguishes between fraud, on the one hand, and honest error, on the other, plus a third category in between, which he labels "murky," where scientists' motives are not "pure" (McIntyre 2019: 134). What McIntyre calls the murky category we call "tweaks." Tweaks are the intentional manipulation of empirical results through changes in and choices of research design, model specification, and/or estimation procedures. McIntyre restricts fraud to data fabrication and manipulation; the "murky" third category does not, in his view, qualify as fraud. Here is why: "What about all of those less-than-above-board research practices p-hacking and cherry-picking data . . . ? Why aren't those considered fraud the minute they are done intentionally? But the relevant question to making a determination of fraud is not just whether those actions are done intentionally, it is whether they also involve fabrication or falsification. . . . The reason that p-hacking isn't normally considered fraud isn't that the person who did it didn't mean to, it's that . . . p-hacking is not quite up to the level of falsifying or fabricating data" (McIntyre 2019: 136).[4] McIntyre clearly stands in the tradition of Robert Merton, who suggests that outright fraud is a rare practice, but also claims that "more frequent practices just beyond the edge of acceptability" exist (Merton 1957: 651). Merton mentions selection and reporting of cases based on outcomes.

"Not quite up to the level" and "just beyond the edge of acceptability" are imprecise phrases that hide more than they reveal. Simple murder does not quite rise to the level of genocide, but that does not make it "not murder." McIntyre's assessment ultimately borders on a tautology: First, he accepts only data fabrication and manipulation as fraud, and then he suggests that tweaks are not quite up to that level. By his definition, that is.

McIntyre demonstrates that one can define fraud in a way that excludes tweaks. However, fraud can also be defined in a way that includes tweaks. Definitions are not right or wrong; they serve a purpose by defining a border, but this border is often fuzzy. The question thus is whether excluding tweaks from the definition of fraud has more advantages than disadvantages. In our view, McIntyre's definition of data fraud is incomplete and imprecise. It conceals that the fabrication and manipulation of data and the influencing of empirical results by means of tweaking serve the same purpose: to promote the researcher's interests.

The term "intentional" is important. In principle, fraud and error are easy to distinguish by definition: Fraud is intentional and error unintentional. And from a statistical point of view, fraud biases result in certain directions, while unintentional error does not bias results. As McIntyre (2019: 133–134) explains: "In the case of mere error, one's fidelity to science is not the issue, for one can make a mistake without the intent to deceive. But in the case of fraud, where any flaws are deliberate, one's commitment to the scientific attitude is squarely in question. . . . By committing fraud, one is putting oneself and one's advancement ahead of this. Ideology, money, ego and self-interest are supposed to take a back-seat to evidence." George and Buyse (2015: 161) agree: "Forgetting to report a value is honest error, deliberately not reporting the value is fraud; incorrectly copying a value is honest error, purposely changing the value is fraud."

Starting from the notion that fraud is intentional, Gall and Maniadis (2019) define fraud as the generation of "false positives":[5] the false rejection of a null hypothesis in scientific terms or the false confirmation of the existence of an effect of a variable or treatment X on the outcome Y. They distinguish fraud from other causes of false positives, notably, errors and omissions. This is a misunderstanding: Manipulated research results are not necessarily false.[6] It is possible to fabricate data that generate estimated effect sizes that are close to the truth. This does not make data fabrication honest or scientific. Data fraud is a practice, not an empirical result.

Both data fabrication and manipulation can be fraudulent but need not be so. Data fabrication is a common and legitimate simulation technique used, for example, to analyze the properties of estimators. The imputation of missing data is also, in more than one respect, "fabrication," and forecasts also depend on fabrication of data by a model. As a consequence, neither data fabrication nor data manipulation is per se fraudulent. What makes these techniques fraudulent is the intention of the fraudsters to manipulate their results. Data fabrication only becomes fraudulent when scientists misrepresent fabricated as observed data.

Similarly, data manipulation occurs during many necessary research steps in between the collection of raw data and the analysis of the final data: Researchers need to decide whether observations belong to the population; they need to identify measurement error and identify possible outliers; and they occasionally need to make mathematical transformations in order to bring observed values in line with an operational definition of a variable or reduce the dimensionality of the data by aggregation across different variables. These practices are not per se fraudulent, but they need to be carefully documented, and the effect these manipulations have on empirical results should be clearly described and analyzed. Data manipulations are fraudulent if and only if they are intentionally used to get the desired results. And they become suspicious if researchers do not carefully document all manipulations of the raw data.

Thus, we do not, at least not fully, concur with common definitions of scientific fraud that dominate the literature. It makes more sense to accept the legal definition of fraud in general, according to which "fraud is an intentionally deceptive action designed to provide the perpetrator with an unlawful gain,"[7] and then apply it to scientific fraud:

> Scientific fraud is an intentionally deceptive manipulation of statistical quantities of interest designed to provide researchers with a benefit.

Are Tweaks Fraudulent?

Authors who have fabricated data dominate the Retraction Watch leaderboard. From a fraudster's perspective, fabricating data has numerous advantages: It is easy, can be done quickly, and requires few skills. Fabricated data can produce the desired empirical results with certainty. What you want is what you get. And fabrication provides fraudulent researchers with data that others do not have and, by definition, cannot have since they do not actually exist.

At the same time, data fabrication has an obvious disadvantage for fraudsters: It can, at least in principle, be detected and, at least to some extent, be proven. Data fabrication will always be a risky strategy since the burden of proof tends to be reversed once a certain threshold of suspicion has been crossed: Once accused of scientific misconduct and confronted with some evidence that raises the suspicion of data fabrication or manipulation, scientists accused of data fraud must usually prove that their data are genuine and have been collected and analyzed in an ethical manner. Failure to demonstrate that data have not been fabricated may lead to the retraction of articles that are only potentially fraudulent.

Preventing data fabrication from detection is not as easy as it might seem. As Simonsohn (2013: 1875) explains: "To undetectably fabricate data is difficult. It requires both (a) a good understanding of the phenomenon being studied (e.g., what measures of a construct tend to look like, which variables they correlate with and by how much) and (b) a good understanding of how sampling error is expected to influence the data (e.g., how much variation and the kind of variation the estimates of interest should exhibit given the observed sample size and design)." Selectively manipulating data points can be more easily concealed.

Fraudsters command a portfolio of alternative techniques that extend far beyond and are more subtle than data fabrication and blatant data manipulation. We call these other techniques, these lesser sins, tweaks. According to scientists' best friend, Wikipedia, tweaking "refers to fine-tuning or adjusting a complex system. . . . Tweaks are any small modifications intended to improve a system." Using a similar logic, we define tweaks as the fine-tuning of research designs or model specifications intended to "improve" empirical results: to increase the level of statistical significance, the effect size, the direction of the effect, the functional form of the effect, and so on—all in the interest of the tweaker. Tweaking is always possible because of researcher degrees of freedom: the ability of researchers to influence the quantity of interest by choices made in the research design and model specification. Tweaks serve the same purpose as data fabrication. They aim to make a scientific manuscript more publishable, thereby generating benefits for the tweaker: more publications, publications in better journals, and more citations, all of which mean a better chance for tenure, promotion, outside offers, salary increases, attention, and fame. Or they bring empirical results in line with the researchers' theoretical and ideological priors. In any case, tweaks serve the interest of the tweaker.

One argument for excluding tweaks from the definition of scientific fraud is that, otherwise, the number of fraudulent articles explodes. While that is true, the definition of an activity has to be independent of the number of cases this definition generates. Once we define scientific data fraud as the intentional manipulation of empirical results, then fabrication and manipulation qualify. But so do tweaks. The means are different, but if the outcome is broadly similar then this does not matter. Perhaps our analogy with murder will help. Murder, according to the legal systems of most countries, is the intentional unlawful killing of another human being. The definition of murder rests on intention, not on the means of murder: Torturing someone to death counts as murder, but so is stabbing one in the back or the deliberate poisoning of the victim's food or drink.

Tweaking meets the definition of scientific fraud. Authors intentionally not admitting to having selected research designs or model specifications that give the desired result misleads readers into overestimating the reliability and validity of the empirical results and the inferences derived from these results. As a consequence, a manuscript might be accepted and published in a better journal and may receive a broader or more enthusiastic reception, which opens the doors to more research funding, better positions, and higher salaries.

Tweaking harms the integrity of science. Just like data fabrication and manipulation, tweaking differs from honest mistakes because it is both intentional and, well, not honest. Tweaks come in many shapes and forms, some of which are known as p-hacking, harking, data dredging, or cooking. McIntyre's "murky" category does not consist of a few outlying cases, few and far in between. Tweaks are, for all we know, much more common than data fabrication and manipulation. Focusing on the latter and neglecting the former makes the problem appear considerably smaller than it actually is. The words scientists like to use for McIntyre's third category are euphemisms, and this also holds for the term "tweak." Tweaks seem almost harmless, a trivial offense, a *cavalier's delict* as the Germans say. But tweakers manipulate scientific empirical results with intent and to further their own interests.

Diederik Stapel is a fraudster with fifty-eight retracted articles under his belt, ranking eighth on the Retraction Watch leaderboard as of August

2024. Stapel came to fame as a fraudster; he has contributed massively to the existential crisis of social psychology. Joel Achenbach, in an article for *The Washington Post*, called him the "lying dutchman."[8] A fraudster he is, but he is surprisingly willing to talk and write about his fraudulent career. He even wrote a book-length manuscript about his life—an autobiography titled "Faking Science: A True Story of Academic Fraud" (Stapel 2014). Whenever we need insights from a fraudster's perspective, Stapel is a good, perhaps the best, source. A Paulus of social psychology that turned Saulus. And then Paulus again, when he started to tell junior researchers that scientific fraud violates the written and unwritten rules of the scientific code of conduct.

Stapel kick-started his fraudulent career by becoming "impatient, overambitious, reckless" (Stapel 2014: iii). Data analyses do not always match the expectations and interests of the researchers. And so Stapel took the truth into his own hands and decided to take "one, tiny little shortcut" (iii). He tortured the data to bring results in line with the arguments of his articles. In his autobiography, Stapel explains how he drifted further and further away from the path of virtue: "Everything had to be neat and orderly. No mess. I opened the computer file with the data that I had entered and changed a . . . 2 into a 4; then . . . I changed a 3 into a 5. I . . . made a few mouse clicks to tell the computer to run the statistical analyses. When I saw the results, the world had become logical again" (103).

In the early stages of his fraudulent career, he eliminated cases that he classified as "deviant"—cases that prevented the results from turning out as he expected and as he wanted them to be. These, in his view, were common practices among social psychologists. Tiny little shortcuts, he calls them. Tweaking. Tweaks were Stapel's gateway drug. Soon after he started to tweak empirical results, he resorted to data fabrication and outright data manipulation.[9] Stapel draws a line in the sand. While he accepts data manipulation and fabrication as fraud, his "tiny little shortcuts" are common practice, and thus not fraud, at least not really. If everyone cheats, is it still cheating? If everyone cheats, isn't it foolish not to cheat?

And that is the main difference from murder. Murder neither is defined by its means nor does murder become a lesser sin when murder rates rise. In contrast, the scientific community counts the deliberate manipulation of empirical results as fraud if the fraudsters fabricate or manipulate data. And not as fraud if they tweak their empirical results. If everybody does it, it can't be fraud, can it?

The scientific community is ill-advised to define fraud by its means or by how common a tweak is. Full stop.

Scientific fraud should be defined as the deliberate manipulation of empirical results in the interest of the authors. This definition clearly includes tweaks. Tweaks are not just tiny little shortcuts; they are tiny little shortcuts with substantively large consequences. Because these tiny little shortcuts change the results of empirical analyses, they make manuscripts more interesting. Manuscripts that become more interesting change reviewers' attitudes toward them, and so tweakers publish in more visible journals and with better publishers. Tweakers publish more interesting results in more visible places, and so they get additional attention for their work. And then they receive better job offers and promotions and they rise to power and influence.

Tweaking is not, at least not primarily, about changing the course of science; tweaking is not primarily about misleading other researchers into investing time and money into hopeless research projects that build on the tweaker's findings. Nor is tweaking, at least not primarily, about the misuse of public research funds, although it is a scandal that hard-working taxpayers fund the research of tweakers. Tweaking is a suboptimal equilibrium in a competitive arena where scientists pursue their interests. And it is about the limits of a competitive system based on trust and on freedom from control by institutions that enforce rules.

Is intentional manipulation of statistical quantities of interest always fraudulent? As with any categorization, it is possible to point toward gray areas. One potential gray area candidate is the experimental researcher who, after a first round of experiments, fails to get the treatment effect to be significant, and so organizes a second round of experiments in the very rational expectation that the sheer number of observations will eventually push the significance level above the threshold that separates publishable from unpublishable results. This research practice is fairly common in the life sciences because experiments are costly and may have unnecessary adverse effects on participants. It may therefore make sense to start with a small sample and only add participants when the results are "not yet significant."

The problem with ever-increasing sample sizes is that as the number of observations approaches infinity, the standard error of an estimate shrinks to zero. Thus, unless the predicted effect given the model specification is exactly zero, all estimated treatment effects will eventually become significant, even if they are trivial in substantive terms. This is primarily a blatant weakness of the p-value as the sole publication criterion and of ignoring substantive effect sizes. Statistical significance should not be accepted as the sole publication criterion—though we understand of course that there is a huge gap between what should be and what is.

However, researchers should be alarmed if their treatment effect becomes significant only with very large sample sizes. Such a weak effect is caused either by a treatment effect that is very close to zero, by a very low signal-to-noise ratio, or by a treatment effect that is rare and exists only in a subsample. Any of these three reasons should preclude publication of the paper before the true causes of the initial insignificance have been identified. In other words, the manuscript should not be published until it has been determined whether an absolutely or relatively small effect or a strong conditioning factor is the cause of the lack of significance.

Scientists may be reluctant to call this practice "fraudulent." However, at the heart of the problem is the overselling of research results and overgeneralization of empirical findings. We agree that this practice is not fraudulent if the article clearly states that the results are insignificant given the original small number of observations and only become significant in the larger sample. But this holds for all adjustments: A change in model specification or research design is not fraudulent if the change and what it does to results are clearly discussed and not suppressed. What makes tweaks fraudulent is not the tweak itself, but the selective reporting of results based on relevant quantities of interest. For example, a gradual increase in sample size is fraudulent if the authors suppress the results of the smaller sample.[10]

Is the researcher aware of this problem, and do they collect more and more data until the desired significance appears? Perhaps not, but as we have said, when it comes to tweaking, it is usually impossible to prove intention. At the same time, the existence of a gray area with manipulations that border on the fraudulent does not mean, for example, that the intentional dropping of a control variable from the list of regressors, adjusting the operationalization of a key variable, dropping cases from the sample, etc., to produce desired results does not constitute scientific fraud.

The Demarcation Line Between Specification Searches and Tweaks

Tweaks challenge science because they are fraudulent *and* difficult to detect. Different research designs and model specifications can be justified, and tweaks are usually indistinguishable from honest specification searches that scientists rely on to improve the specification of their empirical model. Scientific analyses combine observations, data, a research design, a model specification, and a method to make inferences. None of the assumptions researchers necessarily have to make in an empirical analysis are beyond contention. The majority of decisions researchers make can be replaced by different decisions that can be justified and that are no less plausible than the ones actually taken. But different decisions often give different results.

Specification searches are not well covered by methodological textbooks (Bartels 1997). Methodologists prefer to *assume* that researchers know or know how to find the best model specification to recover the underlying true data-generating process. But specification searches are an important part of research. Researchers do not know the true data-generating process and lack the knowledge to closely approximate it. If it were otherwise, research would be a walk in the park, a piece of cake, or, nowadays, the day job of artificial intelligence calculating the few missing parameters.

Elsewhere (Neumayer and Plümper 2017), we have argued that because of the inevitable uncertainty about the typically complex true data-generating process, social scientists should not aspire to develop a model that closely matches it. Instead, based on the principle of parsimony they must trade off simplicity for generality, thereby ignoring many complexities of the real world. Specification searches aim at finding the optimal simplification of the true data-generating process. An optimal model simplifies enough to be tractable, without losing the ability to reliably recover the effect of interest. This is a formidable task, and many specification decisions will remain arbitrary. Specification decisions need to be taken with regard to the population; the properties of the sample; the definition, operationalization, and measurement of the relevant variables; the set of explanatory variables; the complexity of the functional form of the assumed causal relationship between independent and dependent variables; any conditionalities between variables; the temporal dynamics in the data; the spatial dependence between observations; and other factors (Neumayer and Plümper 2017).

Crucially, the need for specification searches does not disappear if researchers adopt robustness testing as a strategy for dealing with the uncertainty about the optimal model specification. Robustness testing examines how estimates and inferences change when one replaces the baseline or benchmark model by plausible alternatives. It has become fairly common that a fifty-page appendix of robustness tests accompanies a twenty-page manuscript. Still, specification searches remain important for specifying the baseline model and for developing plausible alternative specifications.

Specification searches often use statistical information on model fit—the adjusted R^2, the Bayesian or Akaike information criterion, and/or joint F-tests, among other measurements—to select one model specification over alternatives. Model fit indicators are weakly related to the proximity of an estimation model to the true data-generating process. These quantities can be compared if models are nested. Importantly, better model fit does not necessarily represent a better specified model. A model specification that maximizes model fit will potentially overfit the data and lead to a model specification that gives biased results. Other method-driven specification searches rely on econometric tests, for example, the Hausman test, which compares two or more models with each other under the assumption that one model is consistent and the other more efficient. Hausman tests suggest that researchers choose the consistent model if estimates vary, and the more efficient model if estimates are statistically indistinguishable. That may make sense, but a consistent model does not give unbiased results. A third category of specification searches analyzes the structure in the residuals of estimation models, such as the Ramsey specification test for functional form misspecification. A fourth category of specification searches explores whether the data and variables conform to certain assumptions that are vital for a specific estimator.

These tests used in specification searches face many challenges and problems (McAleer 1994). These do not concern us here. Despite misgivings about the usefulness of some specification tests, we have no doubt that researchers must search for specifications and will learn about specifications as part of the process of undertaking their research. Rather, the problem starts when researchers leave the path of honest specification searches and intentionally search for those specifications that produce the desired results. Already in 1995, Chris Chatfield suggested that "40 years ago it may have been true that a single model was typically fitted to a given set of data. Nowadays the increase in computing power has completely changed the way in which statistical analyses are typically carried out (not necessarily for the better!)" (Chatfield 1995: 426). Researchers can estimate thousands and millions of models to find the one or the few that give the desired results, and then they report only those that do.

The main difference between specification searches and tweaks lies in the intention of the researcher, but intentions are not observable and often lie in the eye of the beholder. Tweaks and specification searches can be placed on a continuum ranging from "the researcher skillfully adopts specification searches to find the best baseline model and plausible robustness test models" to "the researcher tweaks the model specification to maximize the probability of getting the paper published in a good journal." Most research undertaken in the real world falls somewhere between these extremes.

Not reporting the empirical results of alternative specifications because they do not support the story the researcher wishes to tell is fraudulent. Not reporting the empirical results of alternative specifications because statistical tests suggest that the model is misspecified or because the residuals show systematic structure is not fraudulent. This distinction is not difficult to make in principle, but it is subtle and usually prevents proving that tweaking has taken place in real cases. Of course, we would prefer to see researchers reporting all model specifications that they have estimated but rejected because of model misspecifications. But we do not think this will happen any time soon, if ever.

Specification Searches, Tweaking, Manipulation, and Fabrication

Specification searches, tweaking, and data manipulation and fabrication describe different strategies. Specification searches influence empirical results as researchers try to improve the research design and model specification. Data fabrication and manipulation generate the desired results. Tweaks sit uncomfortably in the middle: Tweakers change the research design and model specification to influence results in a desired direction.

Specification searches, tweaking, and data fabrication and manipulation all influence empirical results, but scientists take different stances toward these strategies. Specification searches are perceived as a necessary element of the research process, a strategy aimed at improving research designs and model specifications. By sharp contrast, the scientific community regards data fabrication and manipulation as fraud and as a serious breach of scientific integrity that must be sanctioned if detected—scientists found guilty of fraud are "excommunicated from the community of scientists" (McIntyre 2019: 140). Tweakers are usually safe from being sanctioned unless their tweaking is blatant.

Table 1.1
Fabrication, Manipulation, Tweaking, and Specification Searches

	Means	Intention	Sanctions?
Fabrication	By inventing data	To determine empirical results	Yes, when detected
Manipulation	By changing data	To influence empirical results	Yes, when detected
Tweaking	By changing research design and model specification	To influence empirical results	Usually not
Specification searches	By changing research design and model specification	To improve research design and model specification	No

If tweaking intentionally influences results in the interests of those conducting the research, should it be sanctioned? Few scholars like to talk about specification searches in public, let alone in their manuscripts, presumably because they fear being accused of doing something sinister: tweaking. Attitudes toward tweaking fall in between the two extremes: Tweaks are considered "dirty" and few scholars openly admit to tweaking empirical results. But since seemingly everyone does it, tweaks cannot be a serious violation of scientific integrity, it seems. It might be comforting to think that if everyone tweaks, scientific competition is on a level playing field. But the playing field is not level if not everyone manipulates empirical results or if some are more likely to get away with these manipulations than others. Clearly, the propensity for tweaking varies greatly from researcher to researcher. Those who work on multiple projects, who do not have strong ideological priors, who have no previous publications that made a similar argument, who do not urgently need a publication, and/or who do not need immediate success to survive in academia will find it easier to part with an idea or an argument that is not robustly supported by data. From the armchair of a tenured position, researchers may even be willing to bury an entire research project. Others will succumb to the temptation of tweaking.

Data fabrication and manipulation, on the one hand, and the tweaking of empirical results, on the other hand, merely differ in the audacity of the researcher committing fraud. From the fraudster's point of view, the choice between fraudulent strategies constitutes a trade-off—the trade-off between the ease with which the desired results can be obtained, on the one hand,

and the probability of detection, on the other. Data fabrication and manipulation are quick solutions that do not require much work, effort, or superior technical skills. They do, however, carry a greater risk of being discovered.

The opposite holds for tweaks: They require at least an equal amount of effort and often more skill than honestly conducted research. But they are unlikely to be detected. Tweaks cannot be observed but must be inferred. It may well be the case that a scholar's preferred research design or model specification differs markedly from those preferred by other researchers. This might raise the suspicion of tweaking, but suspicion is not proof. Researchers have different perspectives on research design and model specification, and thus results differ. And while whistleblowers have uncovered many cases of data fabrication and manipulation, they have nothing to gain by suggesting that empirical results have been tweaked. Scientists may often suspect tweaking in the work of others, but allegations that estimates have indeed been manipulated remain rare.

The binary categorization of research practices and publications into "honest" and "fraudulent" is therefore a convenient oversimplification. It should be replaced by the recognition that research can be placed on a continuum ranging from specification searches uninfluenced by the interests and ideology of the researcher to outright data manipulation and fabrication. Tweaking, arguably, sits closer to data fabrication than to honest specification searches on that continuum.

The vast majority of researchers are likely to shy away from data manipulation and fabrication. But that does not mean that they shy away from tweaking their results in a desired direction. The main puzzle of scientific fraud is not why some scientists publish fraudulent papers and others do not. The real puzzle is why some scientists resort to data fabrication and manipulation when they also have the option of tweaking the model specification or the research design to get the results they want.

Data fabrication and manipulation are rightly regarded as the most serious forms of data fraud, but they are not science's main challenge. They are a parody of empirical research, a distortion that leaves, with important exceptions, little impact on scientific knowledge and progress. The main problem science faces are tweaks. Tweaks are common, often indistinguishable from specification searches, and either accepted or ignored. Tweaks, because of their sheer number, are likely to have a greater negative impact on scientific integrity than data fabrication and manipulation.

A Line in the Sand

Scientific fraud violates the written and unwritten rules of scientific research. Rules, however, develop their strongest effect when they are clear, when violations can be detected with ease, and when enforcement is simple and not prohibitively expensive.

And here lies the problem with scientific fraud: Scientific fraud is not clearly defined and the gray area is massive. The more broadly we define scientific fraud, the larger the share of fraudulent analyses that are extremely difficult to detect. The more broadly we define scientific fraud, the more costly enforcement becomes. However, if we define it narrowly and exclude tweaks, science will not be able to appropriately address, let alone overcome, its credibility crisis.

Science is ill advised to narrow down the definition of scientific fraud just to make detection less demanding and rule enforcement less costly. The negative consequences of scientific fraud are not limited to data manipulation and fabrication; tweaks have just the same distortive effect on competition for academic merits and research funding, and tweaks have the same devastating effect on public confidence in scientific results and on trust between scientists.[11]

Both scientists and the public lose confidence in science when the probability is not negligible that scientists manipulated the empirical results either to support the arguments, theories, hypotheses, and stories they wish to corroborate or to cast doubt on the arguments, theories, hypotheses, and stories that contradict the worldview they believe in.

Science has lost some of its standing with the public. While skepticism about scientific findings can be healthy and is an inherent part of the scientific process, a general disbelief and distrust of scientific findings pose significant challenges. Scientists have a vested interest in regaining some of that lost trust. This is easier said than done. But much would be gained if scientists were honest about the uncertainties associated with scientific results—honest with other scientists in scientific publications and honest in public statements. Scientists must learn to distinguish between scientific results and their private opinions, and they should promote brutal transparency in scientific research, not hide potential conflicts of interest, and find ways to improve communication between themselves and the public in order to rebuild trust.

2 Fraudsters' Motives

Applause is nice. Success feels good.
—Diederik Stapel

Murphy's Law

After his serial fraud was detected, his papers were retracted, and the University of Leiden had suspended his employment, Diederik Stapel was surprisingly willing to talk about his fraudulent career and to explain his motives and strategies. No other detected fraudster has given nearly as many interviews as Stapel has. No other fraudster has been as frank as he has been about the temptations associated with success in academia.[1]

In an interview with *The New York Times*,[2] Stapel mentions ambition, but, more importantly in his view, he manipulated data because he was "frustrated by the messiness of experimental data." He describes scientific misconduct as an addiction that pulled him deeper and deeper into carrying out acts of increasingly daring fraud. In his response to the interim investigation report of the University of Tilburg, he wrote: "I failed as a scientist. I adapted research data and fabricated research. Not once, but several times, not for a short period, but over a longer period of time. I realize that I shocked and angered my colleagues, because of my behavior. I put my field, social psychology in a bad light. I am ashamed of it and I deeply regret it."[3]

An all too human story, one might say. Anything that can go wrong, will go wrong.[4] If the manipulation of empirical results is an option that generates benefits, somebody will go down that route. Inevitably. Being human, scientists are motivated by interests, the information they hold, their beliefs, attitudes, vanity and rivalry and competition among themselves for scarce resources, jobs, reputation, and fame. No wonder, then, that some turn to data fraud. But scientists are not just typical human beings; they are also members of a professional group who collectively subscribe to scientific integrity, the scientific attitude, and the scientific method. Being thus honor-bound is not compatible with data fraud. True, much of the behavior that is expected of scientists is not formally codified or written down. In the immortal words of Captain Barbossa, in the movie *Pirates of the Caribbean: Curse of the Black Pearl*, scientific integrity, the scientific attitude, and the scientific method are "more what you'd call guidelines than actual rules." But it is clear to every scientist that fabricating and manipulating data are fraudulent.

Quite a few scientists seem to struggle to come to terms with the argument that tweaking squarely falls within the realm of fraud. However, the sheer fact that no tweaker will happily admit that they have influenced their research designs or model specification choices in subtle ways to get to their desired results demonstrates that intuitively scientists know that tweaking is not legitimate either.

What, then, motivates some scientists to fabricate, manipulate, and tweak while others do not? In a detailed review of the experience he gained as a member of various evaluation committees investigating alleged scientific misconduct at the California Institute of Technology, David Goodstein concludes that "three conditions seem to be present when scientific fraud occurred . . . : The three conditions are that the scientist feels himself to be under career pressure, he thinks he knows how the experiment would turn out if it were done properly, and he is working in a field where precise reproducibility is not expected." This account does not answer why other researchers who were under the same intense pressure did not fabricate data and manipulate results. Goodstein also fails to explain why these fraudsters did not stop when they received tenure and the admiration of their peers for their brilliantly invented articles in leading journals.

Fraudsters are, with few exceptions, fairly typical scientists. They tend to be liked by their colleagues, provide public goods to the scientific community, and supervise and collaborate with PhD students. As Sting could have coined it: Fraudsters love their children too.

Moreover, although Japanese scholars are overrepresented on the Retraction Watch leaderboard, known fraudsters have affiliations in virtually all countries and work in different fields, though medicine and psychology are overrepresented.

We see the following five not mutually exclusive narratives arising in the relevant literature that seek to explain why researchers engage in fraudulent behavior:

1. There simply are good and bad people, and thus good and bad scientists, and these bad scientists ("rotten apples") cheat.
2. Identifiable personality traits increase the probability that scientists will engage in fraudulent behavior.
3. Researchers pursue interests—personal, material, ideological, and other interests—and thus resort to fraud when honest research stands in the way.
4. The competitive system of science with its publish or perish logic provides strong incentives for committing fraud and leads to the spread of fraudulent behavior as honest researchers lose out against their fraudulent competitors.
5. The system of science does not have enough control mechanisms and sanctions in place.

These arguments fall into two extremes: On the one hand, there are those who see researchers who commit scientific fraud as rotten apples, as liars and cheats violating behavioral norms who deserve to be thrown out of the paradise of academia. On the other hand, some regard fraudsters not as inherently malevolent, but as individuals of frail moral constitution who, though not absolved of blame, have been driven to misconduct by the relentless pressures of an intensely competitive academic environment. The intense competition between researchers, the publish or perish attitude, the often bizarre and unjustified concentration on just a handful of journals that count—all these factors mount pressure on researchers, and some will ultimately crack.

Once again, the question is whether nurture or nature determines the outcome. As always, both answers can be correct at the same time. The reality of data fraud is more complex than any narrative can account for. Systemic factors help to explain the frequency with which academic fraud occurs and is detected. Personal traits contribute to our understanding of the share of researchers who fall prey to the temptation of data fabrication and manipulation. Competition for scarce resources, for publications, tenure, promotions, and attention, can, in turn, push those researchers with a vague predisposition to cheating over the edge. But even the fiercest competition for scarce resources propels only a minority of scientists to fabricate or blatantly manipulate data. Scientists have a choice. They can engage in data fabrication and manipulation or not, and they can engage in tweaking or not.

Rotten Apples and Individual Character Traits

The simplest answer as to why scientific fraud exists, the answer many academics like best, is that there are black sheep in every large flock. Rotten apples are just the rare deviation from the high moral integrity scientists show—whether this integrity is a natural trait of the majority of scientists, or enforced by rigorous policing (Merton 1957: 651). Fraudulent researchers are just as common (or better: uncommon) as liars and cheats and tax evaders in everyday life. Well, hopefully not as common as tax evaders . . .

But rotten apples do exist. The Paolo Macchiarini scandal provides an extreme case in point. Macchiarini was a visiting professor at Stockholm's Karolinska Institute working on stem cell biology. In 2011 and 2012 he implanted artificial tracheas coated with stem cells in three patients. The patients were terminally ill and the hospital described the operation as a vital indication—necessary to save lives. Two of the three patients died within a few months of the operation, while the third patient died in 2014—after surviving for roughly two and a half years.

In 2013, Karolinska University Hospital stopped further operations with the synthetic trachea and decided to let Macchiarini's contract as a surgeon expire. Macchiarini continued his experimental operations in Krasnodar, Russia. In June 2014, a first allegation of scientific misconduct emerged from a researcher in Belgium. In the following months more and more researchers, including four doctors at Karolinska Institute who had co-authored papers with Macchiarini, claimed that "research results have been described in overly positive terms in seven of Paolo Macchiarini's scientific papers, which . . . incorrectly describe the postoperative status of the patients and the functionality of the implants."[5] The Karolinska Institute launched an investigation and asked Bengt Gerdin to write an external opinion report on potential misconduct. In May 2015, Gerdin concluded that Macchiarini had been guilty of misconduct. According to an article in *The New York Times* written by Henry Fountain a year earlier,[6] Macchiarini had failed to obtain ethical permission to conduct his experiments and published an article that made false claims of clinical success.

This was not the end of the story. Later that year the Karolinska Institute surprisingly cleared Macchiarini of misconduct in a statement that concluded that "Paolo Macchiarini was not guilty of scientific misconduct," though "his research failed to meet the standards of quality set by KI and the scientific community."[7] Karolinska apparently believed that poor research and sheer incompetence are not misconduct. Strangely, the *Lancet* applauded and saw "a remarkable turnaround."[8] And remarkable it was.

The judgment did not last. Almost two years later, Karolinska finally fired Macchiarini; the university's vice chancellor resigned; Karolinska initiated a complete overhaul of its misconduct procedures; and Macchiarini's paper, titled "Experimental Orthotopic Transplantation of a Tissue-Engineered Esophagus in Rats," published in *Nature Communications*, was retracted by Macchiarini's co-authors. In 2022, Macchiarini received a suspended sentence for "causing bodily harm," but not for assault. Overconfidence in one's own beliefs can lead to serious scientific errors, misinterpretations, misconduct, and, as in the case of Macchiarini, physical harm.

It is tempting for science, as a public enterprise that spends taxpayers' money, to regard exposed fraudsters as black sheep and to interpret the retraction of published articles as proof that scientific self-correction works well. But the rotten apples story immediately becomes less convincing when we move beyond data fabrication and manipulation and instead focus on tweaking, the deliberate alteration of research designs and model specifications to produce desired results. Tweaking appears to be more widespread than the "few rotten apples" interpretation of scientific misconduct suggests. In many disciplines, tweaking is part of the scientific culture, a practice that many engage in.

The rotten apple story can therefore only provide part of the picture, but it oversimplifies the social logic of scientific fraud: Most scientists are neither all good nor all bad; they respond to their environment, to competitive pressures, to a severe lack of rule enforcement. And they learn that it is possible to get away with manipulating empirical results. The rotten apples story implicitly suggests that nothing can be done to influence the number of fraudulent empirical analyses—rotten apples are rotten apples—an assumption that we find highly unlikely.

Fraudulent researchers fabricate and manipulate data or tweak empirical results with the same goals in mind, the same utility function, as honest researchers when conducting their research and writing their papers. They seek publications, career opportunities, attention, and fame. These goals are comparatively scarce and thus unlikely to be reached, which triggers intense competition between researchers. And in this competition diligent researchers are likely to lose out when they have to compete with fraudulent researchers (Grimes et al. 2018). Still, the incentive structure of a publish or perish culture and the fierce competition among researchers potentially explain the variation in detected fraud across disciplines. But it does not explain why some researchers become fraudsters and others would rather eat their keyboards than manipulate empirical results.

One explanation is that the former have very different individual characteristics from the latter. It is easy to believe the rather tautological claim that personality determines cheating, but like any tautology, its predictive power is limited at best. Some people choose to cheat by publishing fraudulent articles while others do not. Both fraudsters and honest researchers obviously share the same incentive structures; they operate in the same competitive environment. If this is true, then there is no other explanation for fraud than that the personality of fraudulent researchers differs from the personality of their honest colleagues.

Psychologists have tried to identify the personality traits of cheaters. For example, a group of Dutch psychologists (Tijdink et al. 2016) designed a survey on self-reported cheating among more than 500 "biomedical scientists"—a group made up mainly of students. According to the authors, they found that self-reported research misconduct was associated with a personality trait usually referred to as "Machiavellianism." As reading Machiavelli's most famous book *Il principe* suggests, individuals have Machiavellian traits if they believe it is not wise to tell others one's secrets, that it is better to be successful than completely honest, that it is wise to avoid conflicts with people whom one might need in the future, and that timing matters when it comes to getting back at people (Jones and Paulhus 2014: 31). According to Tijdink et al. (2016: 2), Machiavellianism describes "a person's tendency to be unemotional, detached from conventional morality and hence to deceive and manipulate others, to focus on unmitigated achievement and give high priority to own performances." How this definition relates to the questions asked in the survey we do not know. The authors find that higher scores on Machiavellianism were significantly associated with higher scores on the severity of research misbehavior. There was a trend, marginally statistically significant, for narcissism and psychopathy to have a similar association with research misbehavior. In multivariate analyses that included all three personality traits, only Machiavellianism was significantly associated with misconduct—but keep in mind that the sample of this study is not representative of the scientific community.

These findings need to be taken with more than one grain of salt. The study relies on a rat race between variables that are not independent of each other (survey responses are rarely truly independent of each other), ignores endogeneity problems, and reads a lot into pretty small effects. These effects are significant, but statistical significance does not imply much. As the number of observations approaches infinity, any correlation becomes statistically significant.

The vast majority of fraudulent researchers were respected among their peers before their research misconduct was detected and the world turned its back on them. This suggests that the personality traits of fraudulent researchers are not too different from the population of researchers. More importantly, significant correlations between a Machiavellian personality and self-reported research misconduct should not obscure the fact that the majority of researchers with Machiavellian personality traits do not fabricate or manipulate data. As Gross (2016: 700) summarizes the problems with attempts to explain scientific misconduct in terms of personality traits, "scientists guilty of misconduct have been found in many fields and at different levels in universities and research institutions; their social and educational backgrounds vary. There appear to be no systematic empirical studies of the characteristics of perpetrators of scientific misconduct and no good evidence for any common characteristics."[9]

With this caveat in mind, the individual characteristics of fraudulent researchers seem to consist of the following mélange of various broad character traits that are not mutually exclusive.

First, detected fraudulent researchers often seem to be unable to adequately calculate the risks of their behavior. Again, this borders on a tautology. If fraudsters have been detected, they must have made a mistake that led to their detection. Over time, the chances of getting away with data fabrication and manipulation may have declined as scientific fraud has become a common theme in the media. But nobody knows whether the share of fraudulent research has actually fallen or whether the rate of detection has risen, or both.

Second, data fabrication is the easiest and quickest way to "gather" data and to assemble desired results. It allows researchers to maximize the number of publications. Many fraudsters seem to have become trapped in a vicious circle of wanting ever more publications, attention, and fame. And the easiest way to get this, at least for a limited time, is the fabrication of data that support interesting ideas. And then they publish too much, are too successful to avoid raising suspicion. Scientists seek acknowledgment, recognition, and fame. Even if they have little to gain, or no financial or ideological stake in their research results, they may be tempted to tweak empirical results just to get a boost to their self-esteem by seeing their research published in a good journal or to receive admiration from their peers.

Third, in contrast to fabrication, tweaking empirical results requires methodological skills. Researchers who have sound knowledge of statistical methods tend to be the better tweakers, while data fabrication and manipulation are the means of choice for researchers with limited methodological skills.

Fourth, data fabrication and manipulation give researchers perfect control over results. These strategies are the method of choice for researchers with strong ideological priors on results that have a high probability of being wrong. The lower the chances that simple tweaks will produce the desired results, the more researchers with strong ideological priors must resort to data fabrication and manipulation to get what they want.

The personality traits narrative closely resembles the rotten apples story with one difference: Rotten apples comprise a random share of the population, whereas the personal traits story tries to identify fraudsters. However, both logics have serious problems in correctly identifying fraudsters. Their predictive power remains low. Instead, both logics identify an unacceptably large number of false positives and false negatives.

Biases and Overconfidence

While it is often impossible to demonstrate that single studies suffer from vested interests, considerable evidence exists for bias at the aggregate level. For example, research financed by corporate sponsors is many times more likely to find supportive evidence than research not sponsored by corporations. This holds even when other "sponsorships" are present (Fabbri et al. 2018).

Bias due to interests goes far beyond economic interests, however. As Honeycutt and Jussim (2020: 81) argue, "political bias can slip in and distort the research process and scientific pursuit of truth at many stages, influencing who becomes an academic social scientist, the questions asked, the measures used, how research findings are interpreted, ideas and findings being suppressed, what is cited, and the canonization of research findings." These political biases often become apparent in entire research areas. "When scale labels condemn conservatives, or certain types of studies are systematically ignored or go unpublished, or certain types of questions go unasked, the scientific literature itself becomes politically biased, regardless of whether individual researchers harbor such biases" (Honeycutt and Jussim 2020: 81).

The scientific method was invented to reduce the influence of ideological, political, economic, and other interests on the results of empirical research. There are, however, limits to what the scientific method can accomplish. Researchers have interests, preferences, and beliefs that muddy the scientific waters. Many cases of fraud, and one of the main motives for tweaking, seem to stem from researchers not accepting to be proven wrong by the data. When empirical results do not agree with the theoretical priors and beliefs of the researcher, they often assume that there are issues with the data: Measurement error, model misspecification, and/or faults in the research design all can drive a wedge between the "expected truth" and the empirical results. And thus, they tweak to bring the empirics in line with their "expectations."

If interests and beliefs motivate researchers but reality does not comply, this causes cognitive dissonance. Interested fraudsters reduce the dissonance by fraudulently closing the gap between the results they want and the results they get if they conduct their analysis honestly. Not surprisingly, then, many detected cases of scientific fraud include an element of fabrication for ideological reasons. They were motivated by a deep desire to get a specific result: Joachim Boldt fabricated data to support a medical treatment that, as empirical studies have demonstrated, has no beneficial effects and severe side effects. Andrew Wakefield still to this day seems to be so convinced by his fraudulent ideas that he turned himself from vaccination researcher to anti-vaccination activist. These types of fraud result from incentives structures that go beyond obsessive-compulsive publication disorder. These researchers not only want more publications; they fabricate specific results—results that fit their worldview.

Not all researchers, it seems, will accept evidence if it contradicts their theoretical expectations, their desire to have made important discoveries, their confidence in their own scientific abilities and achievements. Contradicting established scientific truths ultimately leads to major scientific contributions, even paradigmatic changes. Scientific success comes with determination, a willingness to overcome obstacles, including perhaps some tensions between theoretical predictions and empirical evidence. But, as always in life, there can be too much of a good thing. Determination can lead to ignorance and strong confirmation bias (Chambers 2017).

Some exposed fraudsters suffer from overconfidence in results that were at best fragile and at worst manufactured. Overconfidence, broadly defined, refers to a personality trait that causes people to have a subjective assessment of their own knowledge or skill that exceeds the actual level of that knowledge or skill. When scientists are overconfident in their own ideas, they do not allow empirical evidence to challenge those ideas; as Mark Twain suggested, they "never let the truth get in the way of a good story." But there is no source that can be attributed to Twain for this statement. But that is exactly what Twain could have said, so never let the truth get in the way . . .

Just like the personality traits story, the cognitive dissonance story tends to overgeneralize. Data often do not support theoretical ideas, regardless of how strongly one believes that these ideas are correct. The scientific attitude requires that if empirical evidence conflicts with a theoretical prediction, researchers must prioritize the evidence (McIntyre 2019). Researchers are not supposed to torture the data until they confess. Neither should researchers quietly adjust their theory.

Competitive Pressure and Contagion Effects

The majority of fraudulent researchers do not try to explain their behavior. Either they feel there is little left to say or their lawyers tell them not to say anything that could be used against them before judge and jury. The most notable exception was Diederik Stapel, who made two arguments in his public defense. First, he suggested that the checks and controls in academic publishing were not strong enough to deter him from fabricating data, an issue we will address below. Second, he submitted the following: "I did not withstand the pressure to score, to publish, the pressure to get better in time. I wanted too much, too fast. In a system where there are few checks and balances, where people work alone, I took the wrong turn."[10]

According to popular folklore, scientific misconduct has been facilitated, at times directly caused, by intense competitive pressures in the academic world. As in any profession, success drives careers. One's publication record exerts a major impact on job offers, tenure decisions, research grants, attention, and fame. Science is not exactly the "winner takes all" system that exists in professional sport, but it surely is competitive.

But what share of data fraud can be explained by competitive pressure?

In a review of the case of fraudster Michael LaCour, published in *The New York Times* on May 22, 2015, Adam Marcus and Ivan Oransky, the key figures behind the Retraction Watch project, suggested that "science fetishizes the published paper as the ultimate marker of individual productivity. And it doubles down on that bias with a concept known as 'impact factor'—how likely the studies in a given journal are to be referenced by subsequent articles."[11] The argument that fraudsters are the innocent victims of brutal competition comes in many guises and revolves around (1) the fierce competition among scientists, especially between junior scholars for scarce tenured positions; (2) the potentially misleading focus on publishing in a small number of "high-quality journals"; or (3) the intrinsic or extrinsic desire to maximize citations.

The number of media articles blaming fraud on the intense competitive pressure among poor scientists is legion. Here is an example: "The conditions are tough: the research community is globally connected, and the pressure among early-career researchers is extremely high. . . . Today, [academic] careers are decided by the length of the publication list, the reputation of the journals, and the number of citations. But don't we have to pay more attention to the very essence of empirical research—namely, that gaining knowledge cannot be perfectly planned and predicted, but is fraught with uncertainties and risks?"[12]

In this view, academic fraud results from a combination of incentives created by the academic system—the universities and research institutes, the colleagues, the academic peers, the discipline, the journals. Academia provides an environment where research performance can be observed, where a few superstars receive most of the attention, and where institutional control often remains lax. Scientists respond heterogeneously to this environment. The characteristics of researchers, their competitiveness, and ambitions vary along with their knowledge and skills.

The argument that competition for scarce resources in academia causes scientific fraud resembles the idea that income inequality stimulates theft and robbery—a result that remains weak, is shown only in cross-sectional data, and is based on potentially spurious correlations (Bourguignon et al. 2003; Neumayer 2005). Income inequality rarely sparks an increase in property crime, and a decline in inequality does not necessarily lower the crime rate.

The case of Michael LaCour casts doubt on the competitive pressure narrative. LaCour was a grad student at a prestigious institution. And, yes, he sought tenure. But he had already received an offer from Princeton University to continue his work there as an assistant professor. He had already attracted attention for his work, and by no means was he doomed to publish or perish. His academic career seemed to be safe, and yet he fabricated data.

In December 2014 *Science* published an article by LaCour and Donald Green (LaCour and Green 2014) showing that interviewees changed their attitudes about same-sex marriage after talking to a gay person for just twenty minutes. Four months later, *Science* retracted the article on the request of Green and to the fury of LaCour. A lot had happened in a short period of time. Immediately after publication, the finding that attitudes toward gay people are not firmly rooted within individuals but can almost effortlessly be changed by a simple conversation attracted intense media attention from *The New York Times*, *The Wall Street Journal*, *The Washington Post*, and *Huffington Post*, among other outlets. The article also attracted the attention of graduate students David Broockman and Joshua Kalla at Berkeley, who were interested in related research questions and decided to publish an extension of the argument put forth by LaCour and Green. They took a closer look at the data and were immediately irritated by unusually high test-retest reliability rates and equally unusual response and re-interview rates.[13] In their own work, Broockman and Kalla were getting much lower response rates. They contacted the survey company that, according to the LaCour and Green article, had conducted the survey to learn how to keep interviewees on board. Instead, they learned that the survey company had no knowledge of the project and had not collected the data. Broockman and Kalla informed Green, who started to investigate the data, spoke with LaCour, and, as LaCour did not provide the original data, wrote a letter to

Science on May 19 asking for the retraction of the article. This letter was posted on the Retraction Watch webpage.[14]

What makes the competitive pressure argument particularly unappealing is that LaCour and more so Green were both very successful given the stage of their careers. Neither of them immediately needed the publication, though a research article in *Science* is quite an achievement even for successful political scientists. Moreover, LaCour is a junior scholar whereas the majority of detected fraudsters have been senior scholars, often highly respected tenured professors. The Retraction Watch database cannot provide conclusive evidence that most fraudsters are senior scholars, since it samples just detected cases. Nevertheless, the data suggest that most detected fraudulent researchers are senior scholars—researchers with tenure in safe positions who do not need additional publications to survive fierce competition. From this perspective, LaCour seems to be the exception and Stapel the norm: Senior scholars with tenure are more likely to have their papers retracted than junior scholars.

The "competitive pressure" argument paints a misleading picture of the strategies researchers, especially junior researchers, use to thrive in a competitive environment. The best strategy is certainly not to work on a single high-risk project, but it is also not to avoid risky investments in scientific excellence. As the portfolio theory of investment decisions recommends, the superior strategy lies in the combination of rather risky and rather safe projects. This strategy makes survival in a competitive environment possible even when some risky investments fail.

Naturally, the behavior of one scientist partly depends on the behavior of other scientists. The more fraudulent articles are being published, the tighter the competition among scientists for scarce resources becomes. This explains contagion effects of fraudulent behavior and fits well with a narrative popular among discovered fraudsters: Everybody cheats. Their own cheating, their own manipulations, their own fabrications merely level the playing field with other cheaters and fraudsters. In the view of fraudsters, the manipulation of empirical results is an exercise that preserves their own competitiveness in an otherwise unfair competition: "once scientists start cutting corners, that practice has a natural tendency to spread throughout science" (Harris 2017: 187). If fraud chasers try to find the one person guilty of fraud, they need to find the one person who started the fire and invented data fabrication, manipulation, and tweaking. All the others, the latecomers, are merely victims of unfair competition—or so they claim.

Professional cycling and other doping-infested sports have repeatedly relied on this line of defense. Apparently, tweaking, manipulating, and fabricating are to scientists what doping was for cyclists during the era of Lance Armstrong, Marco Pantani, and Jan Ullrich: means to improve competitiveness in an unfair competition with other cheaters, means necessary to level the playing field. As Rudi Pevenage, the manager of Jan Ullrich, explained: "Did we have a choice? After the Festina scandal in 1998, we had stopped everything. . . . No substances, no preparation, nothing! We worked like this until 2001, believing that our rivals were doing the same. . . . We were seeing that Armstrong had become superhuman. What could we do? Let him play while no control could get him? . . . Today we finally can say that we were all the victims of Lance Armstrong."[15]

Dopers are the victims of other dopers, just like scientific fraudsters are the victims of fierce competition in academic publishing. Just like doping, doctoring empirical results resembles a classical collective action problem: Everyone is better off if nobody cheats. Since dopers fare better than honest athletes and fraudsters better than honest scientists, the incentives to cheat loom large. If everyone tweaks, Diederik Stapel suggests, it is a competitive disadvantage not to give in to the pressure of tweaking as well. The honest are the stupid and tweaking is not really fraud anyway, tweakers tell themselves.

Are these excuses plausible?

Evidently, there are advantages to scientific fraud. When fraud works, it improves the publication record of fraudsters relative to their counterfactual selves and relative to honest researchers. With publications come job offers, promotions, and research grants. And since neither the number of full professors nor the available research funding increases as fraud becomes more common, these cheats exert a redistributive effect and induce negative externalities for honest researchers. Still, scientific competition is less competitive than the Tour de France. In science, there is more than one winner. Honest researchers are not doomed to lose the scientific competition for jobs and fame. Scientists can be honest and still survive in a competitive environment. True, scientific fraud gives cheaters an unfair advantage, but it is not necessary to cheat in order to have an academic career.

Insufficient Control and Lack of Sanctions

In an article on scientific fraud published by *Nature*, Jennifer Crocker (2011: 151) concludes that "to understand fraud, we should think about how it begins and escalates, not how it ends. By the time such fraud is exposed, bad choices that would usually lead to only minor transgressions have escalated into outright career-killing behavior." It is unclear what Crocker means by "minor transgressions," but major cases of academic fraud, like the practices employed by those with many retracted articles, often start surprisingly small and then grow into something much larger once fraudulent scientists learn that they get away with minor manipulations. Detected fraudulent researchers frequently suggest that they are the victims of a system that combines powerful incentives to publish with imperfect policies to detect and punish academic fraud. This does not mean that more checks and balances can prevent scientific fraud.

Detected fraudsters merely try to blame others or the system. Just like drivers who drive too fast are not the victims of the powerful engine in their car, fraudulent researchers are not the victims of a system of insufficient control and lack of sanctions. Incentives that motivate drivers to speed and scientists to take research ethics lightly do exist, but no one wants to live in a world that eliminates these adverse incentives through a dense web of intrusive controls and harsh sanctioning mechanisms. Freedom always includes the possibility of violating rules and regulations. That does not make freedom a bad thing.

Scientific fraud is often facilitated by the growing trend toward highly specialized research. As disciplines become more focused and intricate, fewer and fewer researchers possess the broad expertise required to critically evaluate the research designs and statistical models employed. This narrowing of expertise poses a significant challenge (Chevassus-au-Louis 2019), as it becomes increasingly difficult for external experts to assess the rigor and validity of studies. In turn, this trend undermines the quality and reliability of the peer review process, as the pool of qualified reviewers shrinks. The ability to detect manipulated empirical results—whether through selective reporting, data fabrication, or other forms of scientific misconduct—also diminishes. When only a small group of specialists can fully comprehend the intricacies of a particular study, the likelihood of uncovering fraudulent practices decreases substantially. This creates an environment in which flawed or dishonest research can go undetected, compromising the integrity of the scientific enterprise. Therefore, while specialization drives scientific progress, it simultaneously creates significant challenges in maintaining the transparency and accountability of research.

In an interesting article, Smaldino and McElreath (2016: 13) demonstrate that under fairly general assumptions, sloppy and fraudulent research labs will outperform honest and careful labs. Using an evolutionary computational model, they show that "incentives for publication quantity can drive the evolution of poor methodological practices. . . . An incentive structure that rewards publication quantity will, in the absence of countervailing forces, select for methods that produce the greatest number of publishable results." The article also demonstrates that replication studies act as a countervailing force, but, under realistic assumptions, this force remains too weak to prevent the spread of bad research practices.

The Fraud Triangle: Incentive, Opportunity, Rationalization

No single factor can explain why some researchers engage in scientific fraud and others do not. Scientists are a heterogeneous species and that also holds for fraudsters. Andrew Wakefield has little in common with Diederik Stapel other than the fact that both manipulated empirical results because they liked their manipulated results better than the results the data were giving them. The behavior of scientists is determined by a combination of incentive structures and personality traits. Individuals behave differently not only because they are different but also because they find themselves in different situations. The latter results in testable predictions: Fraud will be more frequent in more competitive disciplines and more prevalent in disciplines that have fewer and less effective checks and balances in place. At the same time, researchers who need success for their career are more likely to rely on fraudulent techniques.

The existing evidence for these predictions remains weak. If we accept retractions as an indicator of fraud, fraudulent techniques are most likely to be used in disciplines in which data are collected by the researchers themselves (medicine and social psychology come to mind) rather than using publicly available data, and where institutional checks and balances are few and far between. They are relatively rare among non-tenured scholars, but that may be because the publications of junior scholars are less likely to attract attention. If no one is looking, getting away with fraud is easy.

Data manipulation and fabrication are strongly incentivized by an opportunity structure afforded by privately collected data, data that researchers collect themselves. But the tweaking of empirical results is always an option that researchers have. The possibility of manipulating empirical results does not explain why some cheat and many others do not. Unless researchers fall into the truly rotten apples category, researchers may occasionally fabricate and manipulate data but more likely resort to tweaking empirical results—researchers who have more of the personality traits that make them prone to cheating; researchers who are more driven by ideologies, beliefs, and material interests; or researchers who are too weak to withstand the pressure of the academic publish or perish system. Tweaking usually goes undetected and, if detected, rarely ever results in serious consequences.

3 The Tip of the Tip of the Iceberg

> Scientific misconduct does not happen only at Stanford and Harvard.
> —Ivan Oransky and Adam Marcus

Retractions and Data Fraud

Data fabrication and manipulation differ in their risk of detection. Because data cannot be fabricated unintentionally, data fabrication is the clearest and therefore the most common reason for uncovering scientific fraud. As a consequence, the vast majority of retractions due to data fraud covered by the Retraction Watch database were triggered by evidence of data fabrication. In contrast, the case for alleging fraud becomes less clear when researchers "merely" manipulate data. Researchers may change the values of a few observations in their data or deliberately misread measurement instruments to improve their estimates in desired ways. In fact, cleaning data is not an uncommon step in the research process, and thus data manipulations do not necessarily indicate fraudulent intentions. What matters is intention, and intention is hard to prove. Only if researchers leave a trail of evidence of having replaced values in private data or if they, rather foolishly, change values in public data, is it possible to convincingly prove fraudulent intent in data manipulation.

Fraud cannot be quantified in a reliable way, but survey research suggests that only a minority of scientists fabricate and manipulate data. An even smaller minority of these are caught in action. The majority are unlikely to fabricate or manipulate, unless they do so and get away on a gigantic scale, which is unlikely. Instead, we submit that a large number of scientists engage in tweaking of one form or another. Since the proportion of those who fabricate, manipulate, and tweak but get away with it remains unknown, one cannot know the true magnitude of the problem of data fraud. If only the tip of the tip of the iceberg is visible, the problem is potentially huge.

We do see, however, that the number of detected fraudsters, almost exclusively found to have fabricated or manipulated data, has increased over time. This sounds like bad news. The increase in retractions for data fraud could also be the result of a growing awareness of the problem or more effective detection strategies. If this is the case, the number of fraudsters may actually remain constant or decline. We do not believe in this all too convenient story, but we cannot prove it wrong either.

As the name suggests, Retraction Watch covers the story behind retracted articles. These stories leave the reader with serious doubts about scientific integrity. Retractions occur for different reasons, including reasons unrelated to scientific misconduct. Sloppiness or unintended errors occur (Harris 2017). But the majority of retractions are due to scientific misconduct, and while this term covers more than data fraud, for example, plagiarism, many cases of scientific misconduct involve data fraud.

Data on retractions are difficult to interpret. It would be naive to assume that all fraudulent articles have been retracted. More importantly, with the notable exception of the so-called pizza papers, co-authored by Brian Wansink who had asked some of his students in a pizza restaurant to turn insignificant into significant results, virtually no article has ever been retracted for tweaking. Perhaps the most useful feature of the Retraction Watch page is the intensive coverage of the story behind retractions. We learn about the motivations of exposed fraudsters, often from the whistleblowers who brought them down, less often from the fraudsters. These stories provide insights into the nature and culture of data fraud.

It is not possible, however, to generalize insights from detected fraudsters to not yet detected fraudsters, or from data fabrication and manipulation to tweaking, and we do not know and may never know how common data fabrication, manipulation, and tweaking actually are in an academic environment with some, albeit rather weak, checks and balances in place. The stories about detected fraudsters are not necessarily representative of the population of fraudsters. Some fraudsters decided to mass fabricate fraudulent research papers on such a scale that there was little realistic chance to get away with it. These cheats are overpopulated in any sample of fraudsters—they self-selected into becoming detected. The number of mass producers of fabricated data cannot be known, but there is absolutely no guarantee that the system of checks and balances in scientific publishing works reasonably well to catch even the average fraudster. There are horror stories about universities refusing to accept that a member of their scientific staff has fabricated and manipulated data, of journal editors dragging their heels before they agree to retract an apparently fraudulent article published by their journal, and about universities hiring scientists who have been fired from other universities for scientific fraud.

Impossible that a university would hire a proven fraudster?

Not in Germany in the twenty-first century. Ulrich Lichtenthaler has published numerous articles in leading business journals, received numerous scientific awards, and was featured prominently in an article of the *Handelsblatt*, ironically titled "Der Junge der alles richtig macht" (The boy who does everything right). In 2011 he accepted an offer from the University of Mannheim, one of Germany's leading research universities. A year later, a group of scientists raised concerns about the validity of Lichtenthaler's empirical findings and the integrity of his work.[1] In 2013, he lost his teaching qualification from the Otto-Beisheim School of Management. In 2015, Lichtenthaler resigned from his chair at the University of Mannheim. At the time, fifteen of his articles had been retracted. Three years later, in 2018, Lichtenthaler received an offer for a chair in entrepreneurship from the International School of Management in Cologne—an offer he accepted.

As scandalous as the Lichtenthaler case seems, he will not regain his former glory and fame. He will not return to front-page media coverage or to keynote speeches at prestigious academic conferences. The stories we know support the view that science's ability to correct itself remains intact. And this is the usual story after every retraction, after every detected case of data fraud: The detection of fraudsters shows that the self-healing forces of science are working well, that fraud will eventually be detected, that fraudsters will be expelled from the grace of the scientific community.

How Common Are Data Fabrication, Data Manipulation, and Tweaking?

From the optimistic self-healing perspective, the steadily growing number of articles retracted for data fraud is good news. Every single retraction proves that science knows how to detect and punish its cheaters. Scientists and scientific institutions are paying more attention to the problem of scientific fraud and devoting more resources to the fight against scientific misconduct, and more whistleblowers are coming forward to report potentially fraudulent researchers. Jeffrey Brainard, one of the optimists, suggests that the "burgeoning numbers" of retractions "reflect better oversight, not a crisis in science" (Brainard 2018: 390).

But the pessimists correctly stress that the increasing number of retractions indicates the growing urgency of the problem, regardless of whether fraud increases or not. They believe that more fraudulent articles are detected because data fraud has become more common. However, the credibility crisis of science is fueled not only by fraud but also by the spectacular detections of fraud that make it into the media. Nobel Prize winner in medicine Gregg Semenza, Marc Tessier-Lavigne, and Francesca Gino, three superstars of science, all have had articles retracted for data fraud in just one year, 2023, which has arguably had a large impact on the credibility crisis. Apparently, deterrence has not worked for these three authors.

Scientists often argue that scientific fraud remains a rare occurrence (Steen 2011) and that fraud in science remains an important though not widespread problem. As early as 1957, Robert Merton (1957: 650) concluded that "the annals of science include very few instances of downright fraud although, in the nature of the case, an accurate estimate of frequency is impossible." Uri Simonsohn, a social psychologist and successful fraud hunter, explains: "Outright fraud is somewhat impossible to estimate, because if you're really good at it you wouldn't be detectable. It's like asking how much of our money is fake money—we only catch the really bad fakers, the good fakers we never catch."[2] Absence of evidence, we suggest, is not evidence of absence.

Proof of fraudulent intent is hard to show. Unless fraudsters willingly or unwillingly leave behind evidence for data fabrication or manipulation, proving fraudulent intentions in data analyses will always remain difficult. Attempts to quantify data fraud will inevitably face serious challenges. Data fraud is usually unobservable, and fraudsters are usually unwilling to confess, perhaps not even in surveys that appear to guarantee anonymity. The best existing indicator of data fraud is the number of retracted articles, but (1) not all retractions are based on data fraud, (2) the number of articles that are fraudulent but have not been retracted is presumably very large, and (3) the resources that go into preventing, identifying, and fighting data fraud are not constant over time and space. The number of retractions has been on the rise significantly in recent years. Still, if we consider only confirmed cases, the 65,000 retracted articles that Retraction Watch lists as of July 3, 2025 pale in comparison to the roughly 90 million publications abstracted in Scopus.

How much of the iceberg remains hidden under the surface?

For real icebergs, like those that sunk the Titanic, the hidden part is close to 88 percent, but with data fraud icebergs we don't know how much is hidden. George and Buyse (2015: 162) lament: "Indirect evidence, based on detected cases, will obviously lead to biased (low) estimates, leading to speculation about a 'tip of the iceberg' phenomenon. The under-reporting of fraud may be more severe in medical research than in other scientific fields because of the negative consequences these reports have on pharmaceutical companies or other sponsors associated with fraud. There is incentive for sponsors to divulge as little as possible about fraud or misconduct detected in the trials they sponsor." Malcolm Green, former vice principal of the faculty of medicine at Imperial College London, agrees that the number of detected fraud cases vastly underestimates actual fraud: "It is highly likely that for every case of fraud that is detected there are a dozen or more cases that go undetected. While these might represent lesser degrees of misconduct, they are still important to science and ultimately patient care" (cited in Tavare 2011: 343).

Despite its own limitations, another tentative answer comes from a scientific survey of scientists. A survey found that 8.3 percent of scientists who participated in an anonymous study of questionable research practices at Dutch universities self-reported falsification or fabrication of data (Gopalakrishna 2022: 12). Given that surveys that ask respondents, albeit anonymously, to admit to data fabrication and manipulation are likely to underreport its extent, the 8 percent figure is alarming. But worse, more than half, 51.3 percent, of the survey respondents admitted to "frequently engaging in at least one of 11 'questionable research practices,' which include using inadequate research designs or deliberately judging manuscripts or applications proposals unfairly." Similarly, a US study using 3,247 mid-career scientists as the sample found that 0.3 percent of respondents admitted to having falsified data, 6 percent to suppressing conflicting evidence, and 15.5 percent to adjusting the study design or the research methodology or to changing results in reaction to funder pressure (Grimes et al. 2018).

The results from the Dutch survey lie at the upper end of estimates from other studies. In a meta-analysis of scientific fraud, Fanelli (2009) concludes that a weighted average of 1.97 percent of survey respondents admit to having fabricated, falsified, or manipulated data at least once. However, the statistical uncertainty of that estimate ranges from 0.86 to 4.45 percent. Many more, more than one in three researchers who participated in the survey, admitted to other questionable research practices. In addition, it seems plausible to assume that survey results suffer from underreporting.

Scientists do not like to admit to engaging in questionable research practices, and anonymity does not necessarily make them more honest. Arguably the best indicator of potential underreporting is the gap between responses to the questions "Have you" or "Have your colleagues" engaged in questionable practices. When asked about the behavior of their colleagues, respondents reported on average that up to 72 percent of their colleagues had engaged in questionable research practices (Fanelli 2009). This does not ultimately prove that the majority of scientists have actually engaged in questionable research practices. But these are worryingly high numbers.

Many recent surveys on attitudes of scientists toward scientific misconduct rely on standardized survey questions developed by Kalichman and Friedman (1992: 770) for a "pilot study on perceptions concerning research ethics." The questions asked in this study get directly to the point: "If you would expedite publication of your work, would you be willing to fabricate data or select or omit data to fit your hypothesis?" Only 0.4 percent of the survey respondents admitted to data fabrication, while approximately 7.5 percent answered that they would select data to get the desired results. More recently, Holm and Hofmann (2018) used the Kalichman scale survey to probe research integrity among doctoral medical students at the University of Oslo. Survey respondents are confronted with a battery of statements on scientific misbehavior and asked to give an answer ranging from "completely disagree" (1) to "completely agree" (5). We have reproduced the questionnaire in table 3.1.

Likert scales are extensively used in research on attitudes, and yet methodologists disagree whether these scales work best if they offer an even or an odd number of answers and whether the two most extreme answers ought to be combined into a single category because some respondents are unlikely to choose extreme answers.

Holm and Hofmann find some of the highest average agreement with question 7, which asks respondents whether it is acceptable to selectively omit contradictory results, and question 8, which asks respondents whether it is acceptable to falsify or fabricate data if they are confident of their findings, both undertaken in the interest of expediting publication. On the other end of the spectrum, respondents report high agreement that experimental data fabrication (Q1), intentionally altering experimental results to make

Table 3.1
Kalichman Battery of Research Integrity Questions

Q1	It is never appropriate to report experimental data that have been created without actually having conducted the experiment.
Q2	It is never appropriate to alter experimental data to make an experiment look better than it actually was.
Q3	It is never appropriate to try a variety of different methods of analysis until one is found that yields a result that is statistically significant.
Q4	It is never appropriate to take credit for the words or writing of someone else.
Q5	It is never appropriate to take credit for the data generated by someone else.
Q6	It is never appropriate to take credit for the ideas generated by someone else.
Q7	If you are confident of your findings, it is acceptable to selectively omit contradictory results to expedite publication.
Q8	If you are confident of your findings, it is acceptable to falsify or fabricate data to expedite publication.
Q9	It is more important that data reporting be completely truthful in a publication than in a grant application.
Q10	If you witness someone committing research misconduct, you have an ethical obligation to act.
Q11	If you had witnessed a co-worker or peer committing research misconduct, you would be willing to report that misconduct to a responsible official.
Q12	If you had witnessed a supervisor or principal investigator committing research misconduct, you would be willing to report that misconduct to a responsible official.
Q13	If fabricated data are discovered in a published paper, all co-authors must equally share in the blame.
Q14	If fabricated data are discovered in a published paper, all co-authors must get the same punishment.

Source: Holm and Hofmann 2018: appendix S1.

them look better (Q2), or taking credit for words, data, or ideas generated by others (Q3 to Q5) is never appropriate. No single answer gets close to an average maximum agreement that a specific type of scientific misbehavior is never appropriate.

Survey research is fairly conclusive that the majority of researchers perceive scientific fraud as "never appropriate" and that data fabrication and manipulation count as fraud. However, the Kalichman survey relies on questions that Diederik Stapel would answer without hesitation. Stapel leaves no doubts that he understood that data fabrication is fraudulent—he did it anyway.

Research on violations of research integrity has to be more specific on the "tricks of the trade." At the same time, what is needed is a clear conceptual distinction between attitudes on fraud and fraudulent behavior. Not every fraudster believes that fraud is legitimate, and not everyone who believes that tweaks are legitimate manipulates empirical results.

The set of questions shown in table 3.2 can be used in addition to the Kalichman research integrity questions, but it can also replace them. The punchline of the two questionnaires differs, but these two batteries of questions are certainly related. Whether fraudsters answer honestly to these questions remains, as always, a question no survey can answer.

Table 3.2
Augmented Research Integrity Battery of Survey Questions

Q1	It can be appropriate to only report the significant estimates for a subsample if the estimates for the entire sample turn out to be insignificant.
Q2	It can be appropriate not to report changes to the sample one has made.
Q3	It can be appropriate to test various model specifications to see "what works" and only report these results.
Q4	It can be appropriate to try out different variable definitions, operationalizations and measurements or data transformations to increase the effect size or reduce the standard errors.
Q5	It can be appropriate to drop observations that appear to be outliers.
Q6	It can be appropriate to change selected values for observations that appear to be outliers.
Q7	It can be appropriate to selectively omit contradictory results to let findings appear more robust.
Q8	It can be appropriate to selectively report only the results of robustness tests that do not contradict the main message of the manuscript.
Q9	It can be appropriate to explore different identification strategies to find the largest predicted causal effect.
Q10	It can be appropriate to use a suboptimal research design if its results support the main argument.
Q11	It can be appropriate to include more than one treatment or more than one outcome variable to see which treatment has a significant effect on which outcome.
Q12	It can be appropriate to generalize local treatment effects to the entire population.
Q13	It can be appropriate to reformulate the theoretical prediction based on the results one has found.
Q14	It can be appropriate to implement changes to the model specification even if these changes reduce the validity of the model specification.

Would this expanded battery of questions reveal that a larger proportion of scientists at least occasionally tweak empirical results to improve their manuscripts' chances with reviewers and editors? Perhaps.

The manipulation of empirical results is not a small problem. According to Retraction Watch, only roughly 0.05 percent, or one in 2,000, of the manuscripts published in journals included in citation databases are eventually retracted. This sounds acceptable. But contrast these numbers with Fanelli's (2009) result that approximately 2 percent of scientists admit to having fabricated or manipulated data at least once. Imagine they do so only occasionally, say in one out of four articles, though in reality they probably do so more frequently. But even with one out of four, 0.5 percent of articles would be based on fabricated or manipulated data. This means that on average only one in ten analyses based on fabricated or manipulated data will be detected. Of course, this is a back-of-the-envelope calculation with plenty of uncertainty.

The Detection Paradox

The detection paradox of scientific fraud is that the majority of cases in which authors manipulate empirical results appear to involve falsification; the vast majority—in fact, almost all—of the detected cases of data fraud involve data fabrication and manipulation. While only about 5 percent of researchers admit to fabricating and manipulating data, about 14 percent admit to having falsified empirical results and up to 70 percent to "other questionable research practices." If these figures are accurate, then tweaking accounts for between 75 and 95 percent of the cases of dubious published empirical results. At the same time, even if we define data fraud narrowly, of the more than 3,000 cases in the Retraction Watch database, fewer than fifty articles, or less than 2 percent, have been retracted because of tweaking. Even if these figures are interpreted very cautiously, there can be little doubt that data fabrication and manipulation are vastly overrepresented among retracted articles.

But this is not a paradox. Rather, this imbalance results from two factors: Many scientists do not take tweaking seriously or seriously enough (even though it has very similar effects to data fabrication and manipulation), and it is often impossible to reliably prove tweaking. Both factors lead to an unfortunate situation in which tweakers are almost certain to get away with their manipulations of empirical results. The paradox, if there is one, is that scientists focus on the lesser problem and ignore the more prevalent one.

The relative neglect of tweaking also contrasts with the fact that the replicability crisis and with it the credibility crisis in science have mainly been caused by tweaked empirical results. The replicability crisis had its origins in the medical and life sciences (Prinz et al. 2011; Begley and Ellis 2012; Harris 2017) and in psychology.[3] In psychology, it has been made public when the Open Science Collaboration project (Open Science Collaboration 2015; Klein et al. 2014; Nosek et al. 2022) published an article that replicated 100 widely cited analyses published in three leading social psychology journals. Some 270 psychologists from various universities and research institutes teamed up to replicate prominent research results in psychology. The authors of the replications received no funding, and the authors of the original papers were invited to participate and collaborate in the design of their replication to minimize the potential for error.[4] Almost 90 percent of the replications found a smaller standardized effect than was reported in the original articles. In addition, the correlation between p-values of the replication studies and the original studies does not differ much from 0. Of course, the evidence is not independent: As the effect size goes to zero, the p-value tends to go to one. They found that on average the mean effect size of the original studies halved in the re-analyses; the proportion of statistically significant results declined from 97 percent in the original articles to 36 percent in the replication studies. A measly 47 percent of the estimates of the original estimates were within the confidence interval of the replication studies (Open Science Collaboration 2015).

The impossibility to replicate a single analysis does not prove fraud (Klein et al. 2014). It does not prove tweaking either—at least not for single studies. Effect sizes (and their statistical significance) may well vary over time or a failure to replicate may result from pure chance. However, in the absence of tweaking, replication results should deviate only unsystematically from the original results. But this is not the case here. Results clearly differ systematically, with replication results having smaller effect sizes. In other words, it is not the size of the deviations between original and replication results that indicates potential tweaking by the original authors; it is the direction in the deviations that raises doubts. The most disturbing finding of the Open Science Collaboration project, in our view, was that just under half of the estimates of the original papers were not even contained in the confidence interval of their replication analyses.

Hence, replication analysis of a field of research—like the one conducted by the Open Science Collaboration—is a powerful instrument that provide two pieces of information potentially indicating tweaking: First, the relative size of p-values. If the p-values of the replication are not just different—which is expected—but *systematically lower*, some original studies are likely to have been p-hacked by their authors. And second, any systematically lower smaller effect size in the replication studies suggests evidence for effect size hacking. Let us take a look at how this would show up in replication analyses. Figures 3.1 and 3.2 show identical correlations between two sets of simulated "original effect sizes" and "replication effect sizes." But they allow very different conclusions as to whether the empirical results of the original studies have been tweaked. While the data in figure 3.1 show no signs of tweaking since the differences between original effects and replication effects are stochastic, figure 3.2 displays evidence of tweaking. The effect sizes of the replication study are on average systematically and significantly smaller than the original effect sizes.

There are two important caveats: First, to be conclusive, the number of replicated empirical analyses has to be large to be informative. And second,

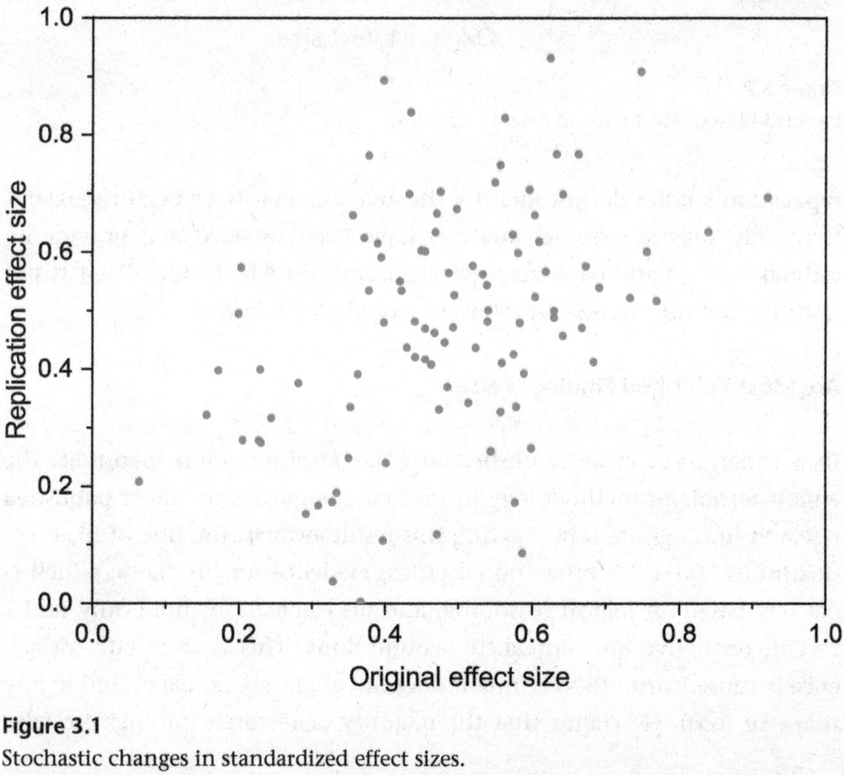

Figure 3.1
Stochastic changes in standardized effect sizes.

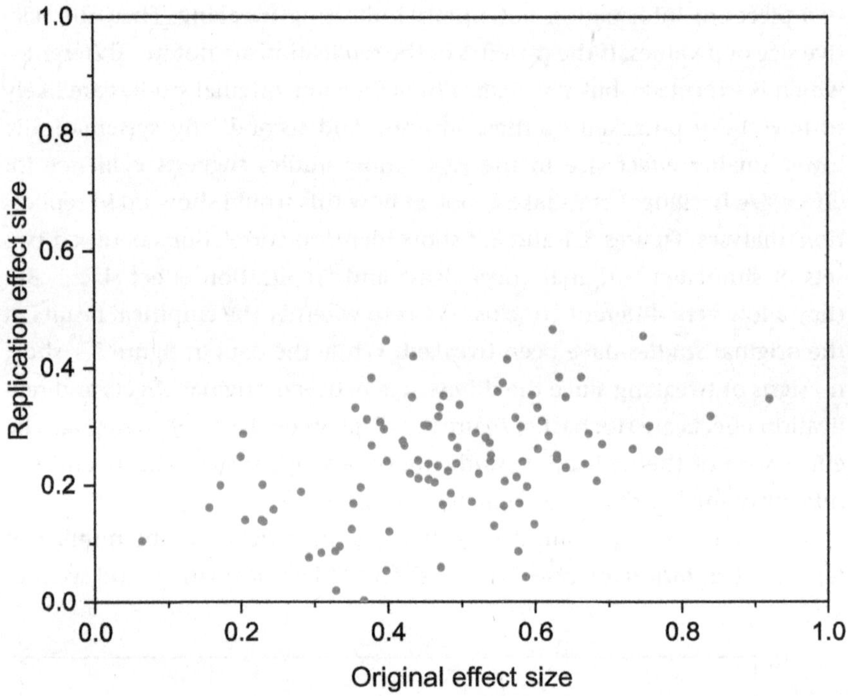

Figure 3.2
Upward-biased effect sizes in original studies.

replication studies do not identify the analyses that have been tweaked—they only suggest whether analyses have been tweaked and provide an estimate how many have been tweaked. But not which ones. That is not nothing, but not necessarily what we would like to know.

Are Most Published Findings False?

In a widely read essay published in *PLOS Medicine,* John Ioannidis, the enfant terrible of methodology in medicine, argued that "most published research findings are false," giving this result away in the title of his essay (Ioannidis 2005). He offers no empirical evidence for his claim, which is entirely based on logical reasoning, and his calculations hold only under certain restrictive and unrealistic assumptions. This is fortunate for science because Ioannidis's argument does not at all rely on data fraud in any shape or form. He claims that the majority of research findings are false even if they are based on honest research. If his assertion is correct, and if many researchers have indeed additionally fabricated or manipulated data or tweaked their results, then the proportion of false published research is significantly higher than what Ioannidis claims.

Fortunately for science, Ioannidis's claim is false—at least false in that it makes sweeping but unfounded general claims about published research findings. Apparently, Ioannidis has learned one or two lessons from YouTube clickbait titles: Make very bold, headline-grabbing claims, whether they stand up to scrutiny or not. Ioannidis's logic goes like this: Many empirical studies are under-powered, their sample size is too small, or the effect, if it exists at all, is too small to recover "true positives" defined as effects that truly exist in reality. Combined with testing hypotheses that are rather unlikely to be true, this can easily lead the majority of published articles with statistically significant findings to misleadingly suggest that an effect exists where there is none.[5] The crucial assumption is that the share of true hypotheses is small—the smaller the better for Ioannidis's claim—while, conversely, the share of false hypotheses is large. Ioannidis does not allow for an in-between category—true for some cases but false for others, for example.

We find Ioannidis's core assumption unrealistic. Although research practices vary from field to field and from area to area, researchers generally tend to test hypotheses that have a fairly high probability of actually being true. Researchers do not, as Ioannidis implicitly assumes, test random hypotheses that have a uniform distribution of actually being true. This is simply a misrepresentation of how scientists work and how they select their projects. Based on well-supported theories and existing evidence, they often test hypotheses that have a high probability of being true. As a consequence, the probability of small p-values is larger, arguably much larger, than Ioannidis assumes.

We have more sympathy with the second assumption, that many empirical studies suffer from low power. However, the main problem of modern empirical research is not the sampling distribution of errors. The main reason to worry whether a large share of published findings is false comes from the potentially widespread practice of tweaking. In Ioannidis's view, the problem is that the p-value threshold is not strict enough.[6] In our view, a stricter p-value will only make tweaking more common. The incentive to tweak will remain as long as journals and the research community give statistical significance an outsized and unmerited, well, significance.

4 Causes of Detection

> The fraudster's greatest liability is the certainty that the fraud is too clever to be detected.
>
> —Louis J. Freeh

Obsessive Compulsive Publication Disorder

The year 2007 was a great one for some. Apple introduced the iPhone, AC Milan won the UEFA Champions League, and Syukuro Manabe and Klaus Hasselmann shared the Nobel Prize in physics for modeling the Earth's climate. 2007 was not a great year for scientific journals and academic publishing. To date, 169 articles published in 2007 have been retracted because of fraud—a small number compared with all empirical articles published in 2007, but a large number compared with the number of retractions of articles published in any other year. In fact, articles published in 2007 account for almost 6 percent of all retractions related to data fraud. The number of articles retracted for data fraud continues to rise over time. In the 1980s and 1990s roughly ten articles were retracted per year. In the 2000s this figure had risen to approximately forty, and it reached an all-time high of seventy-five articles retracted per year for data fraud in the 2010s. From that perspective, the 169 retractions in 2007 seem a lot.

Many of the reports of detected fraudsters are startling. The most notorious fraudsters, those listed on the Retraction Watch leaderboard, appear to have engaged in a quasi-industrial production of fabricated data, as if they were trying to make their actions known to the general public. However, serial fraudsters start with relatively minor actions, such as making a few "minor" changes to the data, removing a few "outliers," or making a seemingly innocuous transformation to a control variable. And then they lose control. Like addicts, they want more. One fraudulent article per year is not enough; the addiction wants three articles, then five, then ten. A few fraudsters have published more than twenty articles in a calendar year, which is not impossible with honest research but also not common. Like any addict, serial fraudsters need an ever-increasing dose over time. They become junkies for publications and for the admiration of their peers.

According to the Retraction Watch leaderboard, five authors, Yoshitaka Fujii, Joachim Boldt, Hironobu Ueshima, Yoshihiro Sato, and Ali Nazari, have had each at least 100 articles retracted.[1] These serial fraudsters were extremely prolific and successful researchers. Many of them were distinguished scientists, well integrated in their academic communities, members or at times presidents of scientific associations. They served on editorial boards of leading journals and were generally liked by their colleagues. Serial killers tend to make good neighbors,[2] and serial fraudsters make good colleagues. And yet their publication records seem too good to be true, and if something looks too good to be true, it may not be true after all.

Jan Hendrik Schön is such a case. In 2000 and 2001 he published six articles in *Nature* and nine articles in *Science*—along with sixty-one other articles and notes in physics journals. Schön was a superstar in the making: He had "demonstrated" that organic crystalline materials could replace silicon in supercomputers. In 2002 the bubble burst, and in 2003 almost all his articles had been retracted. What goes up must come down.

Had Schön's impressive publication record been possible for a young researcher? Schön's publication record looks suspicious, but consider that in 2001 ten authors published more articles in *Science* than Schön, including Harvard's Charles Lieber, who was considered the world's leading chemist at the time. Lieber has more than 1,200 publications with about 170,000 citations to his name. Is that possible without a bit of cheating here and there? None of these articles, to our knowledge, has been retracted. Unfortunately for Lieber, he got into more serious trouble. In 2021, Lieber was convicted of six felonies, including making false statements to the FBI and investigators from the Department of Defense and National Institutes of Health regarding his participation in the Chinese government's Thousand Talents Program as well as for filing fraudulent tax returns.[3]

Jun Iwamoto and Yoshihiro Sato have the same stunningly successful publication pattern as Schön. Iwamoto and Sato co-authored articles for more than a decade. Although first allegations of potential fraud date back to 2013, the majority of the articles were not retracted until February 2022. Concerns about Iwamoto and Sato's research integrity were first raised by Andrew Grey, Alison Avenell, and Mark Bolland. In an interview with Retraction Watch they say that they became suspicious because of the unusual publication activities of Iwamoto and Sato: "We first raised concerns about the integrity of work by these researchers 8 years ago. The researchers were employed at 3 institutions in Japan and 1 in the United States. They published (more than) 300 papers, a mixture of human clinical trials and observational studies, animal trials, and reviews and meta-analyses. Drs Sato and Iwamoto co-authored about 150 papers. Publications rates and workloads were highly implausible."[4] Above all, Grey, Avenell, and Bolland wondered "whether the Japanese scientific community ever questioned how he (Sato) managed to publish more than 200 papers, many of them ambitious studies that would have taken most researchers years to complete" (Kupferschmidt 2018: 637).

The upshot is this: If a publication record appears too good to be true, somebody may take a closer look. Extraordinary publication success is certainly possible with honest research. Nonetheless, it is suspicious.

Whistleblowers

Whistleblowing is an essential part of effective governance (Ting 2008: 249). Whistleblowers reveal information about wrongdoing that would not otherwise be known. Allowing whistleblowers to come forward and protecting them from retaliation has the potential to generate social benefits by constraining the behavior of individuals or institutions and either deterring them from wrongdoing or at least holding them accountable for any wrongdoing.

Whistleblowing incurs costs for the whistleblower, for the person accused by the whistleblower, and for the organization or group in which the whistleblowing occurs. Whistleblowing typically requires the collection and preparation of information and the need to make a compelling argument for the allegations made. In addition, whistleblowing has social costs for the person who blows the whistle. Whistleblowers are often unpopular: Their allegations reduce the social cohesion of a group or organization and sow mistrust. Thus whistleblowers are vulnerable to retaliation by their institution or the individual accused of wrongdoing.

There are obvious costs to the person accused by the whistleblower. If the whistleblower's case is strong, the person's group or institution will usually sanction the accused. If the case is not compelling or if the group or institution refuses to sanction the accused, reputational damage is difficult to avoid. Doubts often remain because, as the saying goes, there is no smoke without fire. The organization or group may also suffer a loss of reputation. Whistleblowing reveals organizational failure in terms of either rules or enforcement. Whistleblowing exposes organizational failures beyond the acts of the individual fraudster, and potentially has adverse effects on trust within the institution, thereby reducing its social capital. These social costs of whistleblowing contribute to the under-provision of the public good whistleblowing.

The costs to whistleblowers who expose scientific misconduct are often substantial. Many whistleblowers describe the experience they make as a "psychological rollercoaster."[5] Retraction Watch has repeatedly reported that the institutions of those accused of scientific misconduct put pressure on whistleblowers. Techniques used by universities include accusing whistleblowers of charges such as "creating a hostile work environment" or "abuse of authority." Other institutions try to drive a wedge between the whistleblower and their colleagues. "In an Australian university, an entire office was given an extra week of holidays except for one researcher who was barred from the extra leave and found out only when she turned up to work on her own. Another university searched an academic's emails to find examples of 'disrespectful comments' about colleagues which were used as justification for misconduct proceedings."[6] But even when the involved institutions respond responsibly, whistleblowers often feel distressed and anxious until the case is finally resolved, which can take years.

The costs to whistleblowers skyrocket when fraudsters sue them. David Sanders, a biologist at Purdue University, dared to criticize another scientist, Carlo M. Croce. Sanders also contacted journals in which Croce had published and tried to convince the editors to retract Croce's work. In 2017 Croce filed a defamation lawsuit against Sanders. The lawsuit took almost three years to come to a conclusion. Croce lost and the judge suggested that Croce had fabricated or manipulated data in at least thirty instances. In July 2024, fifteen of Croce's articles were retracted, twenty-three others were corrected, and five received an "expression of concern." Nevertheless,

Sanders told Retraction Watch that he had experienced dire career consequences, even though his university supported him: "It was conveyed to me that some of my investigatory endeavors, although they were not directed towards articles with authors from my university, were unwelcome. There was a concomitant withdrawal of resources from my laboratory and me."[7]

Whistleblowing has been an important mechanism for detecting scientific fraud. Many of the fraudsters were identified by whistleblowers, which often had been in close, often collaborative working relationship with the fraudster. Gross (2016: 705) similarly suggests that in the majority of cases, "detection of scientific misconduct appears to come from laboratory colleagues: from students, peers, technicians, and supervisors of the miscreant." This is not surprising: To blow the whistle one must have intimate knowledge of the alleged misconduct. And yet whistleblowers have inside knowledge, but they are rarely co-authors. Co-authors share responsibility for the content of a paper, and when the main culprit falls, the co-authors rarely ever appear entirely innocent. At the very least, they appear to have contributed little to the joint paper or not paid enough attention to their co-author's misdeeds. Thus, for co-authors the situation resembles a catch-22: either they appear guilty or they appear careless, and potentially incompetent. Co-authors of fraudsters often trust that all is well—a practice that Donald Green, co-author of fraudster Michael LaCour, describes as "not unusual," because "without some degree of trust one would have to build in so many checks and so much redundancy into the system that nothing would be feasible except at high cost. So there's a cost of ratcheting up the level of mistrust."[8] And so it remains unclear whether co-authors are guilty as well or whether they are the first and foremost victims of the fraudster. Or both, of course. In any case, co-authors often have little to win and some to lose by blowing the whistle.

Keeping fraud under the rug amid the threat of whistleblowing is a difficult task for fraudsters and tweakers (Shamoo and Resnik 2015). They often employ three main strategies to reduce the risk: Fraudulent researchers do not share raw data with collaborators, work alone with data they themselves have collected, or, if co-authorship is the norm in the field, simply invent co-authors who do not exist.

Most of the retracted fraudulent articles had more than one author. According to the American Psychological Association, Diederik Stapel used his influence and prestige to prevent collaborators and research students from participating in the data collection process: "The so-called data collection occurred at secondary schools where Stapel had connections, and occurred under his sole supervision. He enlisted the help of unknown (fictitious) paid research assistants for data collection. After a few weeks, a complete, coded data set would be made available to the collaborator for further analysis. . . . Stapel was described as intimidating and not tolerating questions from students regarding his refusal to have them involved in data collection. There was some discussion among students and junior researchers about his complete control over the data, but . . . it was felt that he was doing them a service."[9]

However, professors usually do not collect the data and research students do not merely assist in writing the paper. Scientific fraud must be concealed by implementing an unusual and strict division of labor, by refusing to share raw data, or by maintaining a monopoly on the steps by which raw data are transformed into final data. Fraudulent researchers find it difficult to keep data collection under control and to preclude any involvement of their co-authors in the collection and preparation of data. If these co-authors do not want to get involved or implement at least some controls, the risk of having their names appear on a retracted article grows. Others will insist on greater data transparency. In these cases, the fraudster must take risks, perhaps sharing fabricated or manipulated data with co-authors, often working night shifts to "perform" lab experiments in splendid isolation and safe from observation.

To avoid being caught out by collaborating researchers, some fraudsters have invented co-authors to let the fraudulent manuscript look like any other co-authored published article in the field. Regarding faking co-authorship, Mario Biagioli sees two advantages for the fraudster: "Fraudsters might risk to attract too much attention to themselves by publishing at an unusually fast clip, which they can do given that it takes much less time to put out fraudulent publications than properly researched ones. Adding fake co-authors helps to make anomalous productivity look more normal." In addition, "If you publish fraudulent claims of a kind that, if properly produced, would require the collaboration of scientists with different skills, it makes sense to make up co-authors with those skills. . . . That may be especially important if people know that you do not possess all the different skills needed to produce the article you just published."[10]

False and Implausible Empirical Results

No fraudulent researcher has better explained how tempting it can be to bring data in line with one's expectations than Diederik Stapel: "I had an idea and then actually you test those by asking people questions, giving them questionnaires or having them do tasks on the computer. That's what I did, but if the results weren't exactly what I expected, I changed the results" (Schröder 2018, translated from German). Clearly, Stapel believed that one should not allow data to get in the way of a good idea. And he is not alone in this view.

Michael LaCour's data fabrication has been interpreted by Adam Marcus and Ivan Oransky as a consequence of his burning ambition to publish in leading journals, and that may well have been one of the many motives for him fabricating data. However, interviews conducted with LaCour's classmates at UCLA suggest that LaCour was "passionate about the issue of gay marriage."[11] Statements like this do not provide conclusive evidence of a political confirmation bias, however, and so we will never know whether LaCour fabricated data on the transmission of support for gay equality via contact with gay canvassers because that was a sexy topic or because he wanted to believe that one can easily reduce anti-gay sentiments over a short conversation with someone who identifies as gay.

Common sense provides a safeguard against nonsensical research results. If a theory seems implausible, it remains implausible even if empirical evidence seemingly backs it up. Implausibility does not imply that a theory is wrong or that the empirical results are fraudulent, but it does raise suspicion. One particularly suspiciously dubious argument claims that women are more likely to dress in reddish colors when they are close to peak fertility. An article by Beall and Tracey (2013) draws its predictions from three more or less questionable assumptions.

First, women are more interested in dating and mating with men when they can get pregnant—most women want babies rather than just sex, it seems. Second, men are more drawn toward women and more likely to date women when they wear reddish colors—apparently, men can be manipulated by weak signals sent by women. And third, women either consciously or instinctively know about this predisposition of men. It follows, at least according to Beall and Tracey (2013: 1838), that "given women's brief fertile window, it may be adaptive for them to dress in a way that increases their sexual attractiveness to men during this period. Women thus may self-adorn in red or pink during this time as part of a broader drive to appear more sexually attractive at peak fertility." One should be skeptical about theories that make heavy use of words like "may," "would," "could," "possibly," and "perhaps." Theories are not about possibilities; they are about the derivation of testable predictions from a set of assumptions—that these predictions may be wrong is an integral part of the game called science.

That is not the main problem with the Beall and Tracey article, however. In *Slate* magazine, Andrew Gelman took issue with their research design and empirical analyses. He made four straightforward methodological points: First, the sample is not representative. Even if inferences were to be valid for the sample at hand, it seems questionable whether the results can be generalized to other cultures. Second, survey responses that require recall, for example, answers to questions such as "When was your last menstrual period?" often have large measurement errors. Note, however, that Gelman does not explain why this measurement error might be systematic and correlated with wearing red. Third, Beall and Tracey defined the "high-conception risk group" as women whose menses had started six to fourteen days earlier—an assumption that deviates from the majority of medical expertise, which suggests that peak fertility occurs ten to seventeen days after the end of menstruation. This assumption moves the "most fertile period" forward by three to four days and could therefore indicate tweaking. But Gelman points to the fact that some of the articles cited by Beall and Tracey make the same assumption. Gelman gives Beall and Tracey the benefit of the doubt. Fourth, and most important for the question of tweaking, the model specification that allows researchers "to find patterns where none likely exist" is the definition of a color in the Beall and Tracey paper. The authors gave survey respondents nine colors to choose from. But they used only eight categories in the final paper, lumping the answer options red and pink together and label the result "reddish." To Gelman, this smells of tweaking as the authors "could easily have chosen red or pink on its own, and of course they also could've chosen other possibilities (for example, lumping all dark colors together and looking for a negative effect. . . . There are lots of degrees of freedom in those choices. . . . They excluded some respondents they could've included and included other people they could've excluded."[12]

Others have similarly questioned these results. Ulrich Schimmack, the author of replicationindex.com, calculated the so-called R-index of the paper. According to Schimmack, the observed median power (the probability of a "true positive," that is, the probability of an actually existing effect being found in a study) in the Beall and Tracey study is 57 percent. They tested their hypothesis in two studies and found significant results in both, giving a success rate of 100 percent. "The discrepancy between the success rate (100%) and the expected rate of significant results (57%) shows the inflated rate of significant results that is expected based on the long-run success rate of 57% (100% – 57% = 43%)." The R-index is then the difference between the observed power in a study and the inflated rate: 57 percent – 43 percent = 14 percent. We are not aware of any Monte Carlo evidence that tests the validity of the R-index and therefore maintain a healthy skepticism about its usefulness. Schimmack does not share our skepticism and concludes, rather, that "a low R-Index reveals that the two studies provide rather weak evidence for the hypothesis and that the generalizability to future studies and real-world scenarios is uncertain. . . . For this reason, Psychological Science should have rejected the article and asked the authors to provide stronger evidence for their hypothesis."[13] Maybe. But Schimmack and his R-index should not be decisive for passing the stern eyes of the reviewers.

Robustness tests could have, would have, and should have helped. And indeed, it did not take long for a replication study to be published (Blake et al. 2017). Perhaps unsurprisingly, results were different: "Results provide some support for a positive trend between the E:P ratio and the probability of wearing red in young women only, but the effect was found only when all non-red colors were grouped into one category in a binary model. We did not find compelling evidence of this effect when using other statistical techniques. More research is needed to test the robustness of the fertility-red hypothesis and identify when and for whom it applies" (Blake et al. 2017: 132). In other words, results are not robust to the use of assumptions that simplify less.

What makes this case interesting is that Beall and Tracey responded to Gelman's *Slate* article. Two of their arguments deserve attention.

> We categorized shirts that were red and pink together because pink is a shade of red; it is light red. The theory we were testing is based on the idea that red and shades of red are associated with sexual interest and attractiveness. . . . Gelman's concern here seems to be that we could have performed these tests prior to making any hypothesis, then come up with a hypothesis post-hoc that best fit the data. While this is a reasonable concern for studies testing hypotheses that are not well formulated, or not based on prior work, it does not make sense in the present case. We conducted these studies with the sole purpose of testing one

> specific hypothesis: that conception risk increases women's tendency to dress in red or pink. This hypothesis emerges quite clearly from the large body of work mentioned above.[14]

This defense is interesting because Beall and Tracey do not claim that they did not tweak the analysis and exploit researcher degrees of freedom in favor of finding confirmatory evidence for their hypothesis. They merely suggest that the allegation of tweaking "does not make sense" because their model specification decision can be justified by the literature. However, good tweakers always justify their model specification by citing some literature—and since there are many different studies in the existing literature, it is always possible to find someone who relied on a similar model specification, estimator, or research design.

The defense did not prove the innocence of the accused—but neither did the prosecutor provide evidence beyond reasonable doubt for the guilt of the accused. And this is the entire point of tweaking: It is easy for anyone to raise doubt about the honesty of a specific piece of empirical research, but proving beyond reasonable doubt that authors have tweaked empirical results is nearly impossible. All that Gelman would probably have been able to show (and what he should have shown) is that the implausible estimates are sensitive to plausible changes to the model specification. For example, changing the assumption regarding when exactly is the most fertile period and estimating separate effects for red and pink probably changes the results and inferences. Beall and Tracey should not have argued that all assumptions are justifiable by prior research, but rather demonstrated that their estimates are robust to the plausible changes suggested by Gelman. This, effectively, turns the question of tweaking into a question of robustness.

Allegations of fraud due to "implausible" research findings are not confined to the social sciences. In 2002, Rusi Taleyarkhan, a nuclear engineer, published a controversial paper on bubble fusion. This process purportedly achieved nuclear fusion by bombarding a container of liquid solvent with strong ultrasonic vibrations, causing the solvent to heat up enough to fuse hydrogen atoms. The study raised skepticism, prompting other researchers to attempt replicating Taleyarkhan's experimental setup. The results were mixed, with most replication efforts failing. However, in 2006, Edward Forringer, while acknowledging his limited expertise in bubble fusion, claimed to have successfully replicated the results in Taleyarkhan's lab at Purdue University. At the time, Purdue chose not to pursue further investigations. However, amid growing doubts and a congressional panel's involvement, Purdue reopened an investigation in 2007, citing several issues that warranted further scrutiny. In 2008, the board concluded that Taleyarkhan had committed "research misconduct" and had "falsified the research record."[15] Nevertheless, he retained his position at Purdue and continued to receive research funding for his work on bubble fusion.

Media and Scholarly Attention

The LaCour case provides an example of an increasingly important avenue for detecting fraudsters. LaCour and Green's (2014) article attracted significant media attention because they addressed such a trendy topic for both social science research and broader public debates. It is therefore not at all surprising that LaCour and Green's work caught the attention of two graduate students at Berkeley, David Broockman and Joshua Kalla. Just like LaCour and Green, they were interested in the logic of political persuasion. LaCour and Green's article was related to their own research, was published in a leading journal, had stunning results, and was widely discussed in the media, including in *The New York Times* and *The Washington Post*. How could one not try to replicate the results? Broockman was immediately struck by the high return rates reported by LaCour and Green. "We weren't getting a great response rate, and we had no idea why."[16] The answer is simple: The data on which the results of LaCour and Green were based were fabricated.

Most scholars agree that generating media and public interest is a good thing. Yet if you are a fraudster or a tweaker, there can be too much of a good thing. Public attention inevitably leads to a higher likelihood of detection. Fraudsters and tweakers who wish to avoid a public debate about how they manhandled their empirical results into existence should therefore not seek the limelight of the media, their seven minutes of fame. Attention can be self-defeating if the results are too novel, too implausible, and, at the same time, too interesting.

The articles on the influence of peak fertility on the choice of women's clothing colors and on the impact of hurricane names on hurricane fatalities report striking but counterintuitive results. That women wear fancy dresses when they are "on the market" seems possible, just like men shave and take a shower when they are on the market, and both men and women wear their most attractive clothes when they have a date. However, that women dress in "reddish" colors when they do not have a date but are merely close to their ovulation day seems implausible. The same holds for the effect of hurricane names on hurricane mortality. Some people associate masculinity with brute force, but does this really mean that these people will relax their evacuation efforts and other safety measures simply because the hurricane was given a feminine name by meteorologists?

Finding evidence for implausible or counterintuitive predictions raises the attention for an article among scientists working in the same field, among peers. If the article appeals to a broader audience, the media may get interested. For example, the "wearing red"' article was featured in *Life Science*, a web-based media outlet that discusses scientific content for a broader audience, but even there, some cautionary points were being made: "More research is needed to replicate the findings in a larger population. The study also did not measure hormone levels, which is a more accurate way to determine peak fertility. . . . The reason behind the new findings is not clear, and future studies are needed to determine whether women wear red or pink around ovulation because they want to look sexy, the researchers said. It could also be that women may wear these colors because they attract attention, and women, consciously or subconsciously, want to bring attention to themselves at a time when they feel sexy, even if they are not necessarily seeking the attention of men, the researchers said."[17] Other, similar websites are less critical, citing only the published study to say that women at high conception risk were three times more likely to wear reddish clothes.[18]

While the "fertile women wear reddish clothes" story attracted some media attention, the media really jumped on the "female hurricane names kill" story. This story was featured in *Science*, *CNN*, *National Geographic*, *Time* magazine, and *USA Today*. Most of these media outlets have uncritically reported the findings. Some, however, added a disclaimer suggesting that the findings might be fragile. *Time* magazine, for example, included the following in its reporting: "This is one of the first studies to make this finding, and the results are only preliminary. Jeff Lazo from the National Center for Atmospheric Research has poured cold water on the study's findings, stating that in the past hurricanes always had feminine names and there are many other reasons hurricanes are deadly besides whether or not people adequately protect themselves."[19]

Replication Failure

If researchers find the results reported by others interesting but implausible, chances are they will try to replicate the research or test its robustness. The replication crisis is a shorthand for the apparent failure of many studies to replicate. The Open Science Collaboration (2015) replicated 100 experimental and observational analyses selected from three leading psychology journals and found a severe lack of replicability.

The structure of discrepancies between original and replication studies provides ample evidence that some original results have been artificially and intentionally manipulated to reduce the statistical ambiguity of findings. To achieve this goal, researchers usually do not need to fabricate or manipulate data. In observational studies, it is often sufficient to tweak model specifications in an unobtrusive way. Experimental data can also be pushed above the commonly accepted level of statistical significance by changing a few values in the data or by dropping a few observations.

The work by the Open Science Collaboration and others demonstrates the depth of the problem. Replication analyses quantify and visualize the extent to which the results of observational studies and experiments are not robust to replication efforts. Lack of robustness will often raise suspicion, but without further evidence it cannot prove tweaking, let alone data fabrication and/or manipulation. Failure to replicate estimation results can be the consequence of any number of reasons other than tweaking or fabrication. The following are some.

1. The research designs of the original articles and the replication analyses differ in small but consequential ways despite the latter's attempt to faithfully replicate the design.
2. The samples differ and the effects across the samples are heterogeneous.
3. In many respects, the world will have changed since the original article was published and that may have changed the effect size.
4. Environmental conditions or an undetected conditioning factor vary between the original and replication studies (or worse: came into play after the original study was conducted).
5. The failure to replicate is due to random error.

And the list goes on. It is also possible that the results in the original study have been tweaked, manipulated, or fabricated.

Replication analyses demonstrate that scientific publications tend to be overconfident in the results they present. In replication studies effect sizes are, on average, systematically smaller and their associated p-values larger than in the original publications. Replication analyses cannot, however, identify the root causes of the lack of replicability and cannot determine whether an individual article is fraudulent or not. Each single paper that fails to replicate deserves the benefit of doubt until proven guilty. Still, failed replications raise alarm bells. What failure to replicate any individual study therefore can do is to raise suspicion further and put it under greater scrutiny for further evidence of guilt. Not a single retracted study has been retracted merely because the results could not be replicated.

Similarly, meta-analyses, the systematic analysis of the reported results of various existing studies, can demonstrate that some published results are severe outliers from the majority of estimates. However, as with replication studies there remains a logical gap between the aggregate evidence that "not all can be right" provided by meta-analysis and evidence that the authors of a single study have tweaked empirical results.

Data Anomalies

Once suspicion has been raised, for example, because of non-replicability of an article that attracted media and scholarly attention, and if the authors of the suspicious article are unlucky, semi-professional fraud hunters like Uri Simonsohn get on their heels and start their forensic work searching for anomalies in the data. The following two case studies illustrate how this avenue can result in the detection of data fraud.

Case 1

A number of authors failed to replicate a 2012 article titled "Signing at the Beginning Makes Ethics Salient and Decreases Dishonest Self-Reports in Comparison to Signing at the End," by Lisa Shu, Nina Mazar, Francesca Gino, and Max H. Bazerman (Shu et al. 2012), published in the *Proceedings of the National Academy of Sciences*. The article relies on several pieces of experimental evidence. In experiment 3, the 13,488 participants, all customers of an auto insurance company, were asked to report the current mileage of cars and had to sign an honesty statement of "I promise that the information I am providing is true." The location of the signature was randomized and appeared either at the top or at the bottom of the form. Shu et al. (2012) claim that participants assigned to the "sign-at-the-top" condition reported driving more than 10 percent more miles than those assigned to the "sign-at-the-bottom" condition compared with the most recent self-reported odometer reading of the car.

Uri Simonsohn, Joe Simmons, and Leif Nelson looked at mileage data from the UK's Department of Transportation, which, as one would expect, showed an almost normal distribution of mileage per year with approximately 10,000 miles on average and a little right skew.[20] The data reported by Shu et al., by contrast, appears to be uniformly distributed. Simonsohn, Simmons, and Nelson infer that the data used in Shu et al. (2012) therefore must have been fabricated.

Here is our take on the data. In figure 4.1, we plot the first reported mileage reading from Shu et al. (2012) against the second reported mileage reading and we add the upper and lower bounds for the difference between the two. The lower bound indicates no difference between both readings, which

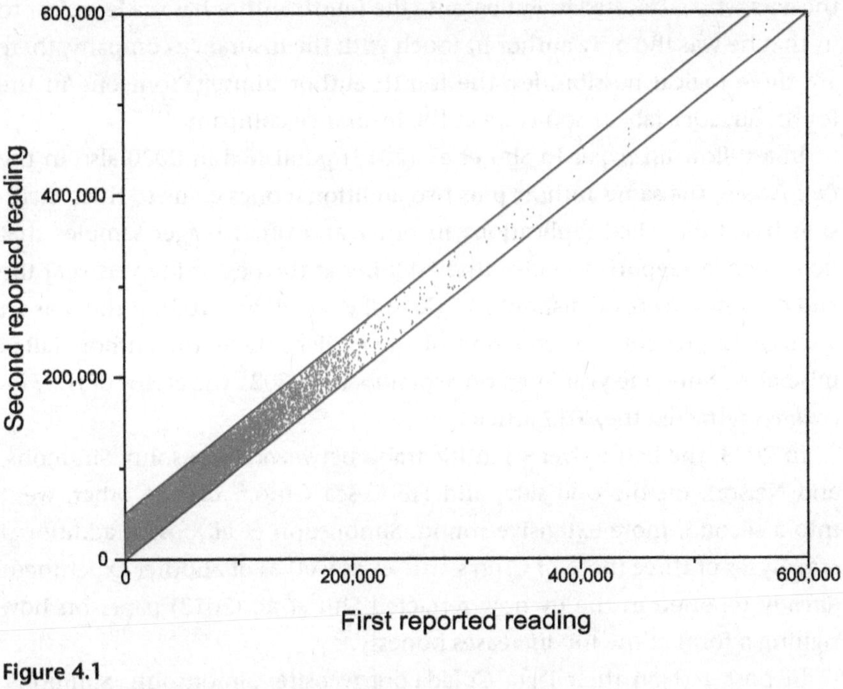

Figure 4.1
Shu et al. (2012) mileage reading (researchbox.org/336).

is possible if the car has not been driven during the time between the two readings. The upper bound is at 50,000 miles, and surprisingly many drivers appear to have driven 50,000 miles between the two readings. This result could have been caused by censoring—or if the data were fabricated by a computer algorithm. The combination of an unusual and unlikely distribution and censoring makes it highly likely that the data have been generated by an algorithm. No car owner has likely been involved in the making of these data.

Simonsohn and his Data Colada colleagues reach the following verdict: "The evidence presented in this post indicates that the data underwent at least two forms of fabrication: (1) many Time 1 data points were duplicated and slightly altered (using a random number generator) to create additional observations, and (2) all of the Time 2 data were created using a random number generator that capped miles driven, the key dependent variable, at 50,000 miles. A single fraudulent dataset almost never provides enough evidence to answer all relevant questions about how that fraud was committed. And this dataset is no exception. First, it is impossible to tell from the data who fabricated it. But because the fourth author has made it clear to us that he was the only author in touch with the insurance company, there are three logical possibilities: the fourth author himself, someone in the fourth author's lab, or someone at the insurance company."[21]

In a follow-up paper to Shu et al. (2012) published in 2020 also in the *Proceedings*, the same authors plus two additional ones come to the conclusion based on failed replications in other and often bigger samples that no evidence supports the idea that "signing at the beginning versus at the end does not decrease dishonesty" (Kristal et al. 2020: 7103). If this was an attempt to prevent the retraction of the earlier article, the authors failed miserably: only one year later, on September 13, 2021 the editor of the *Proceedings* retracted the 2012 article.

In 2023, the battle over scientific fraud between Simonsohn, Simmons, and Nelson, on the one side, and Francesca Gino,[22] on the other, went into a second, more extensive round. Simonsohn et al. posted additional reanalyses of three more of Gino's articles as well as of another experiment already reported in the by now retracted Shu et al. (2012) paper on how signing a form at the top increases honesty.[23]

In post 109 on their Data Colada.org website, Simonsohn, Simmons, and Nelson present evidence that the results of experiment 1 in Shu and co-authors (2012) had been manipulated as well. Experiment 1 was an honesty experiment, where participants first had a chance to cheat on a math test and then on an (experimental) "tax return." On this form, participants reported how much time and money they spent coming to the lab, with these costs partially being reimbursed by the experimenters. The field where they signed that the return was "complete and correct" was located either at the top or at the bottom of the document. Participants thus could lie about their math test and about the financial compensation they claimed and received. Simonsohn and colleagues show (1) that some lines of data look suspicious as they either stem from duplicated row entries or have out-of-sequence identifiers, which suggest that they were tampered with, and (2) that reported results no longer hold when suspicious data points are eliminated.[24]

Figure 4.2 displays the claimed deductions of all participants distinguishing between all treatment groups.[25] We adopt the graphical design and the color coding of Simonsohn et al. where data points are flagged as suspicious (dark) or not (light). If we restrict our analysis to the unsuspicious data points, we find that the deductions claimed by the participants

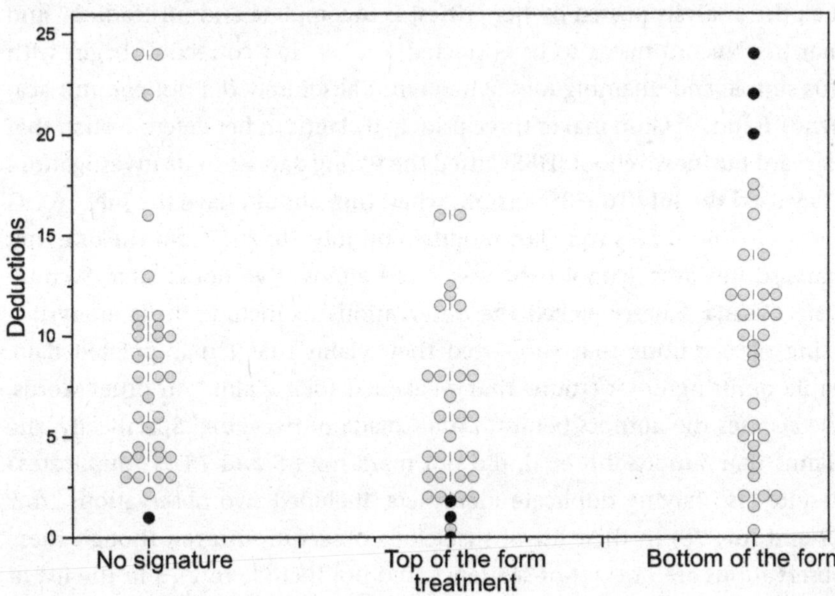

Figure 4.2
Claimed deductions and suspicious data points.

to some extent depend on whether they sign the form or not, but signing at the top does not lead to significantly different behavior than signing at the bottom. It is only the suspicious black data points that appear to drive the significant difference in behavior between signing at the top and signing at the bottom. The suspicious data points are extreme: These presumptively fabricated or manipulated data points suggest that participants claim almost nothing when they do not sign the form or sign at the top, while they claim more than $20 as incurred costs for potential reimbursement by the experimenters when they sign at the bottom.

Suspicious data points that drive results are not conclusive proof of data fabrication or manipulation. However, in post 114, Simonsohn et al. report from court exhibits that were disclosed as part of Gino's lawsuits. Apparently, a forensic firm hired by Harvard University, where Gino is employed but was put on administrative leave in June 2023, found that "52% of reported responses contained entries that were modified without apparent cause" and that many of the values of the dependent variable had been altered so as to support the authors' hypothesis.[26]

In October 2023, Gino started a webpage dubbed Francesca vs. Harvard, which, according to Gino, aims at demonstrating that "the information [i.e., the analysis posted by her critics] is incomplete and misleading" and that the "record needs to be corrected. . . . Let that correction begin with this simple and unambiguous statement: I absolutely did not commit academic fraud."[27] Gino makes three principal claims in her defense: First, that Harvard Business School (HBS) "used the wrong dataset in its investigation. They used the July 16 HBS version, when they should have the July 16 OG version."[28] Both files were last modified on July 16, 2010, but the one that Harvard Business School used was saved almost five hours later. Second, Data Colada "cherry-picked the observations to include in its analysis—using observations that supported their claim that I manipulated data, while omitting observations that weakened their claim." In other words, she accuses the authors behind Data Colada of tweaking! Specifically, she claims that Simonsohn et al. did not mark rows 5 and 75 as "duplicates" despite also having duplicate identifiers; included two observations (row 69 and row 70) in their list of suspicious observations even though these observations are not out-of-sequence; did not include row 73 in the list of suspicious observations though it was also out-of-sequence; and omitted with no good reason the out-of-sequence row 33 from their analysis. Third, "when I excluded all 10 of the flagged observations due to their 'suspicious' nature, and then re-ran the analysis using the remaining observations, the findings of the original study still hold."

Let's dig into the data file and check points 2 and 3.

1. Row 5 and row 75 are not duplicates in the data file we have.
2. Row 68 in our file (69 in Gino's file) is misclassified as out-of-sequence. It is not.
3. Row 69 in our file (70 in Gino's file) is correctly classified as out-of-sequence, unless we assume row 72 in our file is misclassified, then row 69 is correctly classified.

This leaves us with three observations for which it is contested whether they belong into a list of "suspicious observations": row 68, which should not have been excluded by Data Colada, and rows 69 and 72, of which only one should be in the sample.

Table 4.1 reports OLS estimates with "signature at the top" as the excluded category and two included dummy variables for "no signature" and for "signature at the bottom." Note that the treatment effect is represented by the "signature at the bottom" variable. We estimate an intercept that we do not report and we do not include control variables since if

Table 4.1
Gino, Data Colada, and Robustness Estimates of Deductions

	Gino original	Data Colada	Robustness tests: Model 1	Model 2	Model 3
Row 68	In	Out	In	Out	In
Row 69	In	Out	Out	In	In
Row 72	In	In	In	Out	Out
No signature	3.17	2.71	2.87	2.71	2.87
	(1.35)	(1.30)	(1.29)	(1.34)	(1.33)
Bottom	4.35	2.42	2.58	3.06	3.21
	(1.35)	(1.33)	(1.31)	(1.37)	(1.35)
Controls	No	No	No	No	No
N	101	92	93	92	93
adj. R^2	0.085	0.033	0.041	0.043	0.051
F	5.63	2.57	2.96	3.03	3.45

Note: Standard errors in parentheses.

treatment is randomized, controls should not affect the estimates for the treatment effect. We report five results: one based on Gino's original sample, one based on dropping the observations that were deemed suspicious by her Data Colada critics, and three sets of robustness tests that explore various combinations over the three contested observations in rows 68, 69, and 72, but are otherwise based on the sample used by Data Colada.

Results based on both the Data Colada's sample and our robustness tests give estimates that are considerably worse from Gino's perspective than the estimates based on her original sample. The predicted treatment effect is smaller in all replications and declines from 4.35 monetary units based on the original sample to 2.42 monetary units in Data Colada's sample and to 3.21 in the sample that, in accordance with Gino's allegations of "cherry-picking" by her critics, classifies the contested rows 68, 69, and 72 in the way Gino prefers.

Gino claims that obviously she did not manipulate the data because even if one were to accept, which she does not, that some observations are suspicious, once one correctly excludes all supposedly suspicious observations the treatment effect remains significant. Surely, had she fabricated or manipulated the data, then the treatment effect would become insignificant once all "suspicious" observations, and only these, are correctly excluded. And because the treatment effect remains significant with all "suspicious" observations dropped, this demonstrates that she did not cheat. We remain unconvinced: First, the result of robustness test model 3, which follows Gino's perspective on which observations Data Colada should have and should not have included in their analysis is only marginally significant, perhaps too marginal for experimental data to convince reviewers. Second, we do not buy Gino's claim that Simonsohn et al. cherry-picked observations to buttress their claim of data fraud. It is not at all clear to us that, as Gino argues, Simonsohn et al. should have excluded row 72 instead of row 69. One cannot infer from the data which of the two observations has been fabricated. However, row 69 is very similar to the other suspicious data points, whereas row 72 is not. If Gino has manipulated observations, row 69 is the far more probable manipulated data point. Why would anybody manipulate a data point that does not differ from the true data?

In post 110, Simonsohn and colleagues take issue with another set of results of Gino's from experiments conducted at Harvard University and reported in an article titled "The Moral Virtue of Authenticity: How

Inauthenticity Produces Feelings of Immorality and Impurity," published in *Psychological Science*. Again, they find an anomaly in the data: 20 respondents of the survey experiment gave a demographic response that is nonsensical.[29] What makes Simonsohn et al. suspicious that these data points with a nonsensical answer to a demographic question have been tampered with is that they "were especially likely to confirm the authors' hypothesis." In post 114, they report from another exhibit that was disclosed as part of Gino's lawsuits. Apparently, the forensic firm found twenty-five observations in the final data that could not be traced to one of two original Qualtrics survey files. Tellingly, these twenty-five observations influence the results strongly and in the direction of the authors' hypothesis.[30]

Post 111 takes a closer look at experiment 4 reported in Gino and Wiltermuth's (2014) "Evil Genius? How Dishonesty Can Lead to Greater Creativity," also published in *Psychological Science*. In experiment 4, conducted at Harvard University, participants were first experimentally confronted with an incentive and opportunity to cheat on a coin toss. Then participants were asked to conduct two tasks that measured creativity, one of which asked participants to "generate as many creative uses for a newspaper as possible within 1 min" (Gino and Wiltermuth 2014: 976). Evidence for data manipulation results from the fact that Gino and Wiltermuth's dataset has a perfect rank correlation between the row number and a variable labeled "number of responses" that measures creativity. Simonsohn et al. believe that these thirteen data points have been manipulated. And if we replace the creativity score of these observations with one more in line with the others, the effect reported by Gino and Wiltermuth is far from statistically significant.[31]

Finally, post 112 reanalyzes Gino, Kouchaki, and Casciaro's (2020) results from an article titled "Why Connect? Moral Consequences of Networking with a Promotion or Prevention Focus" published in the *Journal of Personality and Social Psychology*. Simonsohn et al. suspect data manipulation because for many observations, the rating respondents allegedly gave on how dirty or tainted a networking event made them feel showed absolutely no variation across seven survey questions. In addition, the ratings often fail to correspond to the words respondents used to describe their feelings. No proof of data manipulation, but certainly a little smelly.[32] More damningly, the forensic firm commissioned by Harvard University found 168 entries in which data points have been altered in the estimation dataset compared with the original Qualtrics survey source data.[33] In a final report Harvard University found Gino guilty of research misconduct.[34] Harvard University responded to this series of reanalyses by requesting a retraction of the articles. In late July 2023, Gino countered the allegations of research misconduct by filing a lawsuit against Harvard University and Simonsohn, Simmons, and Nelson "demanding apologies and at least $25 million." Gino's defamation claim was dismissed in September 2024. A year later, in August 2025, Harvard in turn filed a case against Gino.

Case 2

Dirk Smeesters, a social psychologist at Erasmus University in Rotterdam, conducted research on the influence of the environment on consumer choices. In one of his most cited papers, Smeesters and his co-author Jia Liu purport to study how color can alter the way that primed constructs affect behavior, claiming that "compared to the color white, blue is more likely to lead to assimilative shifts in behavior, whereas red is more likely to lead to contrastive changes in behavior. In our experiment, previous findings were replicated in the white color condition, participants' behavior assimilated to primed stereotypes of (un)intelligence and contrasted away from primed exemplars of (un)intelligence. However, in the blue color condition, participants' behavior assimilated to the primed constructs, whereas in the red color condition, participants' behavior contrasted away from the primed constructs, irrespective of whether the primed constructs were stereotypes or exemplars" (Smeesters and Liu 2011: 653). The problem with these experimental results was that, as Smeesters admitted, he had "massaged" the data. The fraud was discovered and made public by Uri Simonsohn. He replicated the research and data analysis by Smeesters and Liu and became suspicious of "how perfect many of Smeesters' results seemed when, statistically speaking, there should have been more variation in his measurements."[35] In March 2013, the article was retracted. Erasmus University had established that the data used in the article "is fraudulent and invalid" and that the fraud was committed by Smeesters.

The detection of Smeesters' data fraud came at the worst possible point in time for social psychologists. A few months earlier, journals had begun retracting Stapel's articles. And with both working in social psychology, the detection of Stapel's and Smeesters's fraudulent articles pushed the discipline into an existential crisis. *The Guardian* describes this crisis well: "The field of psychology has come under particular scrutiny because many results in the scientific literature defy replication by other researchers. Critics say it is too easy to publish psychology papers which rely on sample sizes that are too small or to publish only those results that support a favored hypothesis. Outright fraud is just a small part of that problem, but high-profile examples have exposed a grayer area of bad or lazy scientific practice that many had preferred to brush under the carpet."[36]

Strategies for Avoiding Detection

Fraudsters can actively reduce the likelihood of detection by adopting a number of strategies, including cutting research collaborations to a bare minimum and, instead of working with real co-authors, naming invented co-authors on the paper to let a manuscript look more "typical" given the preponderance of co-authorship in many disciplines.[37] We have already discussed invented co-authors as a strategy to eliminate the risk of co-authors blowing the whistle.

Not all fraudsters go as far as to co-author their articles with a guy named Stronzo Bestiale—check the translation yourself if you are not familiar with Italian. The story behind the invention of invented co-authorships goes like this: In 1987 William G. Hoover, a physicist at the Lawrence Livermore Institute, received two rejections from prestigious journals for a manuscript on molecular dynamics. In a perhaps somewhat childish reaction, Hoover added Stronzo Bestiale to the list of co-authors and submitted the paper to the *Journal of Statistical Physics*, which accepted it. And so Stronzo Bestiale was born (Penders and Shaw 2020). In Scopus, Stronzo Bestiale appears as being affiliated with the Institute for Advanced Studies in Palermo, an institute that does not host a webpage. Stronzo has not published for a while. We are concerned.

Our friend Stronzo did not stay alone for long. According to Fessi,[38] a second ghost was invented at the Max Planck Institute for Physics in Munich: Prof. Alois Kabelschacht. Kabelschacht started his life as room 354, which had a little tag at the door that read "Kabelschacht," the German term for cable duct. And that it was: a cable duct with a door. A wisecracker working at the research institute one day promoted room Kabelschacht to Professor Alois Kabelschacht, room number 354. As Kabelschacht had obtained his first academic merits, staff added him as co-author to their papers. Kabelschacht's first publication was with Peter Breitenlohner in *Nuclear Physics B*. In 1987, Alois Kabelschacht managed to get his first single-authored paper, on a method now used in the programming language LaTeX. How he did that, we do not know. We have no information whether Kabelschacht receives a salary from the Max-Planck-Gesellschaft.

All fraudsters, whether they have real or only invented collaborators, face a trade-off between maximizing attention and recognition and minimizing the potential for detection and suspicion. Clearly, everything else being equal, fraudsters wish to boost the contribution, originality, and impact of their work, often by challenging "existing wisdom." But this generates attention that may lead to public debate and possibly detection, and that is a big risk for the fraudulent researcher. Yet refraining from fabricating or tweaking results that challenge existing findings is not a winning strategy either since fraudulent researchers seek attention for their results. Even if the fraudulent researcher is the first to enter a new field of research, the success of the fraudulent publications will almost inevitably attract the attention of other researchers. Invisible scientific fraud without any impact is to some extent self-defeating: If it does not attract attention, what is the point? But if it attracts attention, detection becomes more likely. Catch-22.

Despite its pointlessness, one strategy to reduce the potential for detection is therefore to keep the contribution on the small side. In an interview with *The New York Times,* Stapel explained how he was able to get away for so long: "I didn't do strange stuff, I never said let's do an experiment to show that the earth is flat. I always checked—this may be by a cunning manipulative mind—that the experiment was reasonable, that it followed from the research that had come before, that it was just this extra step that everybody was waiting for. . . . So that it was believable and could be argued that this was the only logical thing you would find. Everybody wants you to be novel and creative, but you also need to be truthful and likely."[39] Truthful and likely? In whose judgment? One man's treasure is another man's trash.

Few papers that add little to the existing body of scientific knowledge have attracted attention. The most straightforward strategy for getting away with fabricating, manipulating, or tweaking is therefore to mirror existing research results. But it remains unclear why one would need to resort to fraud for that. Take tweaking as an example, since it is the fraud strategy most difficult to detect. Shadow tweaking, the strategy of tweaking without attracting attention, can be a viable strategy for researchers who are interested in publications for the sake of publications, and potentially for job offers. But when researchers tweak to generate attention and fame, to maximize lifetime income, or to make empirical results conform with ideological priors, shadow tweaking is not a promising strategy.

It helps fraudsters to stay away from politicized and highly controversial research areas or from research areas where scientific progress is likely to lead to commercial success. In general, the more controversial and politicized or commercially attractive the research areas, the greater the likelihood that economic competitors or researchers with competing worldviews will subject empirical results to replication studies and robustness tests.

At least in the social sciences, researchers may get away with fraud if their results are consistent with the ideological biases of their peers. Diederik Stapel seems to have explicitly relied on this strategy: The vast majority of his research findings were interesting and consistent with what we presume to be the social and political biases of most social psychologists. In three of his most cited papers, Stapel claimed that power increases hypocrisy (Lammers, Stapel, and Galinsky 2010), that self-evaluative comparison is more likely undertaken if self-related cognitions are cognitively accessible to the individual (Stapel and Koomen 2001), and that awareness of the "collective we" improves performance on math tests (Marx, Stapel, and Muller 2005). In other words: individual power is bad and a social orientation is good. These messages are perfectly in line with the worldview that dominates social psychology and explain why Stapel's fabricated research results were cited 920 times in 2011 according to Google Scholar. They have continued to be cited well over 200 times per year since his articles were retracted. Arguably, Stapel's research record based on many fabricated and retracted papers still attracts more attention than that of the average article in social psychology.

The strategy of fabricating data that produce results that peers like may have helped Stapel's publication record, his citations, and his reputation and rise to fame. It did not protect him from internal whistleblowers, however. Collaborating with co-authors increases the risk of detection for data fabricators and manipulators and, to a lesser extent, for tweakers. At the same time, the vast majority of manuscripts are co-authored. As a consequence, fraudsters attract suspicion if they predominantly publish single-authored articles. It is therefore not surprising that researchers with a large number of retractions tend to have co-authors. Yoshitaka Fujii, longtime figurehead of the Retraction Watch leaderboard, had at least one co-author and often four or five co-authors on retracted papers. He was caught by a suspicious editor and a statistician, who calculated the odds of Fujii's empirical findings. Joachim Boldt, who took over from Fujii in July 2023 as the one with most retracted papers, published eight of them alone, but he co-authored most of his papers with four, often five other authors. He was brought down by the combined effort of a journal in which he had published and his home university, which ultimately revoked his title for failure to fulfill his teaching obligations. Hironobu Ueshima published most of his articles with Hiroshi Otake, the chairman of the anesthesiology department. His fraud was uncovered after John Loadsman raised suspicion about Ueshima's work in March 2020 because he had smelled data fabrication. Much of Yoshihiro Sato's work, rank four on the Retraction Watch leaderboard, was co-authored by Munetsugu Kikuyama and Kotaro Oizumi. The average of his papers has at least four authors, and many of the papers have more.

The most promising strategy to get away with fraud is to stay well clear of data fabrication and manipulation and instead rely exclusively on tweaking and particularly on those tweaks that are unlikely to arouse suspicion. For example, in a regression analysis of observational data, clever tweakers do not drop important and influential control variables. It is better to add a control variable that gives the desired effect for the main variable of interest or to alter the covariance of a control variable with both the dependent variable and the variable of interest by, for example, taking the logarithm of the control variable. Tweakers can exploit methodological biases of their peers to tweak empirical results. For example, the obsession of economists with instrumental variables invites tweakers to influence empirical results through their choice of instruments. Instrumental variable regressions estimate partial effects, and since more than one potential instrument exists, these models give researchers the choice of an instrument and thereby the choice of a partial effect that the authors do not necessarily acknowledge as a partial effect and instead sell their results as "the true causal effect" of *x* on *y*.

Other tweaks that can masquerade as clever modeling strategies include the choice of an estimator, the modeling of temporal dynamics, spatial dependence, the inclusion or exclusion of unit or group-wise fixed effects, and other techniques. The trick is to tweak a model by making it more complex because models that add bias by complicating the model are less likely to be scrutinized than models that do so by simplifying the model. Skillful researchers can disguise their tweaks by making them indistinguishable from seemingly necessary or appropriate complications to the model specification.

The last two or three decades have seen an explosion in sophisticated research designs, model specifications, and estimators. In the absence of conclusive tests that guide researchers' choices, the potential for tweaking has been amplified by the proliferation of seemingly clever research designs: The more choices researchers have, the greater the potential for tweaking.

Undetectable Fraud

The number of detected fraudulent articles says little about the quantity and sophistication of scientific fraud. They say nothing about the difficulty of finding sufficient evidence for fraud. The majority of detected cases appear to be blatantly fraudulent. It is unclear, however, how representative these cases are for the entire population of fraudsters. Maybe the majority of fraudulent authors are more sophisticated in their choice of fabrication, manipulation, or tweaking than those who have been detected. Detected fraudsters may be different not only from the average scientists but also from the average fraudster. Not every fraudster gets away with it, but getting away with fraud is definitely possible. One cannot know how big the iceberg really is.

The detection of fraudsters and the retraction of fraudulent scientific articles have strengthened the belief in some that existing checks and balances are working. This interpretation may be wishful thinking. The system of checks and balances can only work if either the detection rate or the deterrence rate, the share of potential fraudsters who refrain from committing fraud, is high. Both assumptions seem unlikely.

The effectiveness of deterrence is further diminished because fraud does not just result from quasi-monetary incentives and the desire for fame. Fraud may also be triggered by an ideological preference for a certain result. As fraudsters have claimed, organic materials can replace silicon in supercomputers, bacteria can be engineered that transform biomass into biofuels, traditional Chinese medicine can diagnose and treat COVID-19, anti-gay sentiment can be mitigated by having a conversation with homosexuals. These invented findings have attracted a lot of attention, but it may well be the case that the fraudsters were convinced by their own idea and did not want to let reality get in their idea's way.

A more compelling optimistic perspective holds that if fraudulent research results differ significantly from honest research results, the fraud will eventually crystallize. This logic assumes that the vast majority of scientists are honest. As long as this is the case, the few rotten apples in the basket will be exposed whenever honest research produces different results. This leads to a simple conclusion about the nature of fraud in science: If fraudsters wish to remain undetected, they cannot produce fraudulent results that differ much from the results of other researchers in the field. Undetected or undetectable fraud has thus potentially little effect on results. However, even if fraudulent articles do not influence the course of science, they still inappropriately and unfairly redistribute resources such as positions, research grants, and fame away from honest to fraudulent researchers. The optimistic view breaks down if the share of articles that does not report honest results increases. If deviations from the truth are not random, but systematic, and the share of articles that deviate from the truth becomes large, the self-healing forces of science are seriously undermined.

The real threat to scientific knowledge, therefore, stems not from data fabrication and manipulation but from tweaking. Tweaks are more difficult, and often impossible, to detect than outright data fraud. Tweaking is more common than data fabrication and manipulation, and although each tweaked result damages the fabric of science less than analyses of fabricated or manipulated data, the cumulative damage to science caused by tweaking could easily exceed that of data fabrication and manipulation.

Summing Up the Main Causes of Detection

Cases of data fraud are discovered because whistleblowers report them or because other researchers find the published results questionable and begin a detailed investigation that leads to detection. But this does not explain why a published article becomes suspicious in the first place. Often, the detection of scientific fraud has been facilitated by the behavior of fraudulent researchers. They publish too many articles in too many good journals in too short a time. Academic success is addictive. Or they suddenly publish many more articles per year than they had published before they learned that publishing manuscripts is much easier if they manipulate the empirical results. Other fraudsters are too convinced of a single idea, an idea that they want to be right but that is not supported by empirical evidence.

The implications are clear: If researchers want to become successful fraudsters and still avoid detection, they should not be too successful, should not reach for the stars, and should definitely conduct non-fraudulent research in addition to allay the suspicion of others. Detection becomes extremely unlikely if fraudulent results are not too far from mainstream findings. This successful fraud strategy, while promising, does not necessarily satisfy ideologues among the fraudsters. But it may be only a matter of time before those are discovered.

The majority of fraudsters get away with it. And those who do not get away often make mistakes in their publication strategy that expose them to a high probability of detection: often hair-raising mistakes. The vast majority of detections of fraudulent articles come through one or more of six channels, and some of them are relevant for the less common allegations of tweaking as well:

1. Fraudulent researchers are too successful, which raises suspicion.
2. Fraud is identified by whistleblowers who have insights into the research process.
3. Fraudulent or tweaked research findings report false but eye-catching results or are based on hypotheses regarded as highly implausible by many.
4. Often as a result, some fraudulent or tweaked research generates media attention, which in turn draws the attention of other researchers, including those outside the authors' immediate peer group.
5. Once suspicion has been raised for any one of the four reasons stated above, other researchers will often try to replicate the suspicious study, and if that fails to produce the same or very similar results, suspicion grows further.
6. Semi-professional fraud chasers have started to dedicate their time and effort into detecting fraudsters, using sophisticated techniques to identify patterns in data, figures, pictures, and results that are unlikely to be generated by natural processes.

The detection probability of tweaks differs from that of data fabrication and manipulation. Tweaking is often impossible to detect, so that tentative evidence usually comes from the overrepresentation of statistically significant results that are close to the threshold of insignificance at conventional levels. It cannot be proven beyond reasonable doubt by a lack of robustness of the empirical results nor by a failure to replicate the original results based on new data.

II Researcher Degrees of Freedom

Why is everyone opposed to p-hacking? It is brilliant. Literally anything can be true. We can control the economy. We can be telepathic. We can cure cancer.
—SMBC Comics

Researchers enjoy creative freedom over how they conduct empirical research and how they arrive at their conclusions. This freedom has been dubbed "researcher degrees of freedom" by Simmons, Nelson, and Simonsohn (2011: 1359). Researchers have leeway over the choice of a research question, the theoretical perspective, and, most importantly, over the research design and empirical model specification. They thereby influence the results and findings produced by their research and exercise additional considerable freedom over the interpretation of their results and the inferences they derive from them.

And like all freedoms, it can be exploited. The influence over empirical results that researchers exert with tweaking falls short of the control that comes with data manipulation and fabrication, but tweaks usually suffice to turn unpublishable into publishable results.

The first law of data analysis is that whatever results researchers want to produce, there exists at least one research design and one model specification, and often more than one, that will produce the desired results. Combine this with the fact that tweaks are not frowned upon in equal measure and are often undetectable, and it becomes clear that the real threat to scientific knowledge stems not from data fabrication and manipulation but from tweaking. Tweaking may have a smaller impact on the empirical results in any single paper, but the cumulative damage to science caused by the collective sum of tweaking undertaken by researchers may well far exceed that caused by data fabrication and manipulation.

Researcher degrees of freedom exist regardless of research design. This is the second law of data analysis: No research design, no research strategy prevents tweaking. All empirical results can be tweaked intentionally in the interest of the tweaker. While some research designs and some methodologies offer more freedom than others, there remains enormous leeway in all designs.

The third law of data analysis is that the optimal tweaking strategy depends on the research design. The most promising strategy, the one that is most promising for the tweaker, exploits the particular methodological weaknesses of each specific research design. Researcher degrees of freedom exist in traditional regression analyses of observational data, but equally in experimental research designs as well as in causal models.

Researcher degrees of freedom result from the fact that researchers inevitably have to make choices: about the research design, including the sample on which the analysis is based, and the model specification. These choices are not innocent; they influence the empirical results. Accordingly, researcher degrees of freedom describe the extent to which researchers can deliberately alter empirical results. Of course, researchers are not free to choose any research design and any model specification they want. They are bound by methodology, by the research designs chosen by other researchers who work on similar topics, by methodological conventions and fashions. But the choices left within these constraints usually still give researchers a large influence on empirical results. Knowledgeable and skilled researchers will describe, explain, and justify their choices in ways that appear at least plausible to readers, and that's what they are trained to do. This is the fourth law of data analysis: There are clearly unjustifiable research designs and model specifications, but there exist enough model specifications and research designs that can be justified as plausible by a tweaker.

That the results of all empirical research can be tweaked does not mean that all empirical results have been tweaked. But a clear understanding of the tweakability of all empirical research is the prerequisite for understanding the urgency and importance of countering fraud and tweaks.

5 Tweaking in Observational Data Analyses

Go on a fishing expedition for something—anything—interesting. No, this is not immoral.

—Daryl J. Bem

The Dirty Methodology

1. It is easy to lie with statistics.
2. If you torture the data long enough, it will confess.
3. Lies, damned lies, and statistics.

These are three classic verdicts about the pitfalls of empirical analyses that found their way into folk wisdom. The first quote is credited to Frederick Mosteller, founding father of Harvard's Department of Statistics. The second quote has several origins but is often traced back to Ronald Coase, winner of the Nobel Memorial Prize in Economics in 1991. And the last quote is famously found in Mark Twain's autobiography, although Twain quotes Disraeli here, the former prime minister of the United Kingdom. Twain is usually misinterpreted here: He makes no statement about statistics per se but claims that "figures often beguile him" about his daily output that he measures by the number of words.

The quantity of interest in regression analysis is often, but not always,[1] the level of statistical significance, the p-value.[2] The p-value is without doubt the most controversial quantity of interest in empirical research, since it does not measure what empirical researchers wish to measure:[3] whether a cause has a sufficiently large and sufficiently general effect on the outcome. Perhaps more importantly, many empirical researchers and even methodologists misinterpret the p-value.[4]

Though the p-value[5] dominates discussions about the manipulation of empirical results, any quantity that influences reviewer decisions can become the target of tweakers. For example, if an empirical analysis has an extremely low R^2, a measure of the proportion of the total variance in the dependent variable that is explained by the model, reviewers might reject the manuscript on the grounds that the variable of interest and the control variables do not sufficiently explain the variance of the dependent variable. Other reviewers might ask for the calculation of substantive effect sizes and reject a manuscript because the effect found appears substantively irrelevant.[6] The explained variance or the substantive effect size can therefore become the variable of interest, with tweakers searching for model specifications that make the effects and the explanatory power of a model bigger.

The dominance of p-hacking over the manipulation of other statistical quantities results from two factors: On the one hand, p-values are the quantities of interest on which reviewers often focus.[7] It is not impossible but highly unlikely that an analysis with insignificant findings survives the review process and is accepted by the reviewers and the editor. On the other hand, p-values can be manipulated with ease: Small changes to the model specification often have large effects on the p-value. Andrew Gelman, a methodologist who speaks out clearly when he presumes that model specifications have been tweaked, has demonstrated in a joint paper with Hal Stern (Gelman and Stern 2006) that minor changes in the level of significance—say, from a p-value of 0.045 to one of 0.055—are rather meaningless or, as the authors formulate in the title of their article, "not itself statistically significant." More importantly, Gelman and Stern show that "small and non-significant changes in the underlying variables" can trigger "large changes in significance levels" (328). The criterion that determines the life or death of a paper, the criterion that decides whether months of hard work yield some payoff, not only relies on an arbitrary threshold that cuts off "significant" from "non-significant"; the value itself can easily be manipulated. To make matters worse, in regression analyses, the p-value is an estimated quantity of interest, one that depends on the sampling error. Whether the p-value is 0.04 or 0.06 often depends on mere chance.[8]

Why Results from Observational Data Analysis Can Be Tweaked

In observational data analysis, the results of an empirical analysis depend in part on arbitrary decisions about research design and model specification.

By "arbitrary" we do not mean that the decisions have no foundation or reasoning behind them. Instead, we mean that other choices can be made that peers consider at least as plausible. Scientific theories cannot tell empirical researchers everything they need to know about empirical research design and model specification choices (Freedman 1991).

This is no coincidence: Theories necessarily simplify reality. They usually explain a single causal mechanism and assume validity under ceteris paribus conditions. Empirical analyses of observational data cannot establish such ceteris paribus conditions. Estimated effects are unbiased if and only if all confounders—variables that have an independent or conditioning influence on the outcome and are correlated with the variable of interest—are included as control variables into the empirical model. Because theories do not specify the ceteris paribus conditions under which they claim validity, they offer insufficient guidance to researchers on which control variables to include in their empirical model. Thus, a complete set of control variables derived from theory does not exist in the social sciences.

The choice of control variables is by no means the only modeling decision left at a researcher's discretion, but it has generated the most interest (Leamer 1978; Leamer 1985; Chambers 2017). However, there is more, much more, that can be tweaked. Theories often do not specify the relevant population, the functional form of any effects, or the best measures for a theoretical construct that is not directly observable and measurable. In addition, effect sizes are usually heterogeneous and vary across cases, space, and time. The list could go on but the consequence of model uncertainty, uncertainty about the best specification of an empirical model, is always the same: researcher degrees of freedom in action.

Needless to say, the knowledge to perfectly specify models does not exist. It also remains questionable whether complex specifications outperform simpler specifications. As Freedman (1991) has argued, social science theories do not aim at explaining any true data-generating process. Instead, they focus on one or at most a few causal mechanisms. In order to do so, they have to simplify and simplifications cannot be entirely accurate.

Model uncertainty presents researchers developing an empirical model with choices: some well-founded, some arbitrary, some reflecting the researchers' honest attempts to find the best specification, some opportunistically pushing the quantity of interest in the desired direction. Analyses of observational data are and always will be vulnerable to tweaking. Model uncertainty opens the floodgates to tweakers, making the tweaking of observational data analyses dead easy, thus undermining confidence in them: "Statistical inference can become highly dependent on seemingly trivial but indefensible modeling assumptions. Choosing one or only a few specifications to publish in the presence of model dependence makes conclusions largely nonempirical and subject to criticism from others who make different choices and find contradictory results. . . . When researchers ignore the uncertainty due to model dependence, scholarly works tend to have the flavor of merely showing that it is possible to find results consistent with ex ante hypotheses, rather than demonstrating their veracity" (King and Zeng 2007: 231). Researchers can, if they wish to do so, influence the quantity of interest of an observational study in the "right" direction, the direction that maximizes the publication chances of the manuscript. Seemingly small and plausible changes to the model specification can have large effects on the quantity of interest. The set of plausible or at least seemingly plausible model specifications is huge.

Sampling decisions are a powerful weapon in the arsenal of tweakers. Researchers can change the research design by manipulating the sample, exploiting uncertainty about the boundary between cases that belong and those that do not belong to the population. The definition of a population influences the sample and, indirectly, the inferences being made. And whenever the sampling rule is not perfectly random—each case from the population has an identical ex ante probability of being drawn into the sample—selection occurs, because cases will be over-, others under-represented.

Observational data analyses struggle to adjust for selection and self-selection into or out of treatment. Individuals select themselves into types of education; companies choose a certain ownership and organizational structure; and countries adopt an electoral system based on historical experiences. In all these cases, treatment and control groups are not identical, and treatment status is obviously not randomized across units. Sometimes, conditional homogeneity can be achieved by inclusion of control variables and potentially matching with them. More often than not, it cannot be achieved. Selection and self-selection into the sample or into or out of the treatment group pose formidable challenges for analysts of observational data, which they try to tackle with various sampling, stratification, and selection modeling techniques. But formidable challenges equal formidable opportunities for tweakers.

Researchers can add more cases to their sample or drop some of the cases, calling them "outliers," a practice known as data trimming. The p-value has the property that the variance of the sampling distribution tends to decrease as the sample size increases if everything else is equal. If researchers can add observations to their estimates at will, they can, within limits, choose the p-value they want. A particularly pernicious form of tweaking occurs when researchers keep collecting and adding additional cases until they get the desired result and then stop. This matters more for experiments than for observational data analyses, where the available observational data often limit the size of the sample.

Researchers may restrict the sample to a set of cases convenient or beneficial for the researcher's analysis. Dropping cases reduces the sample size but may still push the p-value under the significance threshold if cases are dropped selectively and with p-values in mind. Samples can be restricted to a particular region or country or time period. Alternatively, one may arbitrarily set a minimum value on a selected variable below or above which cases are excluded from the sample. There exists almost always a subset of cases that allows researchers to find significant support for the tested hypothesis. Researchers rarely concede that their conclusions hold at best for their restricted sample and more or less openly claim to have demonstrated the existence of a globally valid "effect." Selection based on estimation results not only biases the estimates, but also strongly reduces and often eliminates the external validity of the findings.

One issue is how to deal with cases where there are missing values for one or more variables. Values are typically missing not at random. Instead, both observable and unobservable factors influence the ex ante probability that a value will be missing. Researchers can make an effort to find the missing values in another data source or leave them out. They can impute missing values in multiple ways via linear or nonlinear interpolation or extrapolation, theory-based out-of-sample predictions, and multiple imputation techniques, or leave the missing values missing. Estimation results will vary depending on which strategy researchers choose giving them leeway over the results.

Model specification choices provide tweakers with a legion of opportunities. In Neumayer and Plümper (2017) we explore how model uncertainty affects specification choices. Specification choices under uncertainty leave room for tweaks. Tweakable specification choices include the definition, operationalization, and measurement of variables; the set of explanatory variables; the functional form of variables; causal heterogeneity and conditionality; temporal heterogeneity; and heterogeneity in effect strength over time and across space. Honest researchers can test the robustness of their baseline model to plausible alternative model specifications, and in Neumayer and Plümper (2017) we propose more than fifty robustness tests, and this list is certainly not exhaustive. To the tweaker's delight, every one of these tests provides them with a hint on how they can potentially tweak their results. The more robustness tests honest researchers come up with, the more they fill the arsenal of tweakers. This is not an argument against robustness tests but rather highlights how well-meaning methodological advances can be exploited to let results appear less contestable. Only unskilled tweakers will make model specification choices that are regarded as highly implausible by their peers. More skilled tweakers make choices that others may find more plausible. But a mere gut feeling that some specification choices have been deliberately chosen based on previously observed empirical results is no proof of tweaking.

One model specification choice that is promising for tweakers is to alter how the dependent variable or key explanatory variables are defined, measured, and operationalized. We have already encountered creative manipulation in many of the potentially fraudulent articles we have discussed, such as Beall and Tracey's (2013) definition of "reddish" that lumps both red and pink outfits into a single category. We certainly do not mind simplifications, but not if the empirical results depend on them.

But the problem goes deeper. Many of the variables used to analyze observational data in the social sciences are unobservable: happiness, political power, and fairness, to name just three. For these concepts, there is no unit of measurement, no scale, and no measuring instrument to speak of. Contrast that with a watt, the unit of power transmission, where one watt is equal to one joule per second. A watt is a measurable quantity: It comes with a scale, a ratio scale as two watts are twice one watt, and precise measurement instruments exist. And while a watt is a watt no matter where we measure it, one person's happiness category 4 is another person's category 3 and a third person's happiness category 5, simply because social scientists measure (if that is the correct term) on a subjective scale. And whatever the answer is depends on the social context, or perhaps on whether the sun is shining or not.

Books From California
51 W. Easy St
Simi Valley, CA 93065
UNITED STATES
customerservice@booksfromca.com

Books From California
51 W. Easy St
Simi Valley, CA 93065
UNITED STATES

To: Alibris APEX DC 76901949-75 - APEX
800 Avondale Ave.

Grandview Heights, OH 43212-3473
UNITED STATES

If this order is shipping from the US to an international buyer, remember to attach proper documentation. Complete and attach USPS Form CN-22 for all Priority Mail International Envelopes or First-Class Mail International shipments (available at your post office).

Order Number:	33958619
Ship Method:	Standard
Customer Name:	Alibris APEX DC 76901949-75 - APEX
Order Date:	9/10/2026
Alibris Order #:	76901949-75
Email:	

Items:

We are not lamenting the way modern social sciences quantify variables such as happiness. We only wish to stress that how they measure unobservable multidimensional subjective phenomena provides researchers with ample opportunities to tweak their empirical results. Multiple competing and equally plausible techniques reduce multidimensional concepts into a single dimension, while subjective quantifications are heterogeneous and therefore leave ample opportunity to manipulate empirical results through re-sampling. The collection of survey data by researchers themselves gives them access to "private" raw data that only they know and have access to. As the case of Diederik Stapel has demonstrated, a lot can be achieved by turning a 2 into a 3 or a 3 into a 4, let alone by turning a 3 into a 5 and a 2 into a 4, and if data are private, it is virtually impossible to prove data manipulation.

Examples of how questionable sampling decisions together with other questionable model specification choices create suspicion of tweaking in observational data analyses are plentiful, and we pick a couple of them here. In a series of articles dating back to the 1970s, sociologist David Philipps suggested that famous people are able to prolong their lives at least by a few days, for example, to die after their own birthday or other important days such as US presidential elections rather than before (Philipps and Feldman 1973; Philipps and King 1988; Philipps and Smith 1990). Why only famous people? Philipps suggested that birthdays are a repeated reminder that life is continuously moving toward death and are not necessarily occasions for festivities. For famous people, however, birthdays are important public celebrations of their life that they do not want to miss. At the same time, potential tweakers can define fame in numerous ways. By choosing the right sample of famous people one can find evidence in favor of the death postponement hypothesis. And if one wants to reject the hypothesis, there are plenty of opportunities to define fame in a way that gives the desired result.

According to Gary Smith (2014: 179), who carefully examined the data, Philipps coded all deaths that occurred in the birth months as having occurred after the birthday, regardless of whether the birthday was on the thirtieth and the death happened on the first or vice versa. Smith identified twenty-six famous people in Philipps's sample who died during their birth months. Of these, thirteen died before their birthday, twelve after their birthday, and one on his birthday (Smith 2014: 179). This does not give an impressive pattern and certainly not a significant result for the death prolongation hypothesis. Smith suggests that Philipps tweaked until he got the results he wanted. He calls it the "Texas sharpshooter" approach: "Sooner or later, something will work" (Smith 2014: 182).

Carmen Reinhardt and Kenneth Rogoff were already famous economists, perhaps potential Nobel Prize candidates, when they published their politically most influential article in the *American Economic Review Papers and Proceedings* in 2010. In this article, Reinhardt and Rogoff (2010) argue that public debt is detrimental to economic growth but miraculously only above a threshold of about 90 percent of gross domestic product (GDP). The group of countries with debt above this limit experienced a growth rate that was on average 4 percentage points lower than that of the countries below the threshold. The median growth rate of these highly indebted countries was only 1 percentage point below that of the control group. The 90 percent threshold is Reinhardt and Rogoff's main message, a simple message derived from a complex, bidirectional relation between debt and growth that in all likelihood is conditional on other factors.

Reinhardt and Rogoff did not publish their data upon publication but made it available upon request later. Using their data, Thomas Herndon, Michael Ash, and Robert Pollin (2014) identified three major flaws with their analysis; each of these increased the detrimental effect of debt on growth. One of these was an honest coding error in the spreadsheet, later acknowledged by Reinhardt and Rogoff. The second one relates to how country group averages are computed. The authors took the average of the average growth experience of each country in a group where countries were grouped according to the public debt to GDP ratio. That is unusual. Normally, researchers take the average of the growth rates of all countries in a group. Still, the operationalization can, in principle at least, be justified on the basis of concern about within-country serial correlation. We will therefore concentrate on an impactful sampling decision. Reinhardt and Rogoff excluded Australia and Canada from 1946 to 1950 and New Zealand from 1946 to 1949. During these years, the countries were highly indebted but experienced high growth. At the same time, Reinhardt and Rogoff kept the United States during the period included, though the United States had high debt and low growth rates (in fact, negative ones for three out of four years). This looks like selection on results. The exclusion of all years immediately after World War II from the sample could have been justifiable. Shortly after major wars, debt levels are high; the economy recovers from the war depression and might therefore be strong, and this situation can hardly be generalized to periods in which no war occurred. However, selectively excluding some specific postwar country years with high debt and high growth rates from the sample raises suspicion.

Maybe this is tweaking, maybe not.[9] What speaks against Reinhardt and Rogoff is that one honest coding error, one unusual way of calculating average growth performance, and one dubious sampling decision all influence results in the same direction. They also did not publish their data immediately on publication, though replication policies had already become a fairly standard practice in 2010, when Reinhardt and Rogoff published their paper. What speaks for them is that they shared their full data three years after the publication of their article. Better late than never.

Harking

Harking is an acronym that stands for "hypothesizing after results are known." This practice is also known as fishing, data grubbing, data pillaging, and other terms. Such a tweak is simple, but can be very effective: Collect data, estimate a large number of models, identify and report a few nice significant results, and tell a story explaining why the correlation found in selected models is causal. After all, scientists are trained to justify everything; the most bizarre correlations can certainly be argued to be causal.

Instead of testing a predefined hypothesis, harkers estimate results and afterward formulate hypotheses that fit these results and ready they are.[10] Harking amounts to fraud if authors present their inductive results as empirical support for a deductive hypothesis. Harking cannot be detected: No reader can tell whether the chicken or the egg, the theoretical explanation or the significant empirical result, came first.

The rise of powerful computers and large datasets have reduced the costs of harking. As Smith (2014: 4–5) has explained: "Decades ago, when data were scarce and computers non-existent, researchers worked hard to gather data and thought carefully before spending hours, even days on painstaking calculations. Now with data so plentiful, researchers often spend too little time distinguishing between good data and rubbish, between sound analysis and junk science." Essentially, harking is a form of data mining: searching the data for statistical associations and telling a story that fits the mined results. Not everything one can see in data is real, let alone causal. Data are full of spurious correlations, of patterns that exist for a short moment in time. And big data in particular guarantees one can find significant correlations due to the sheer size of the estimation sample, which produces very low p-values.

Harking is not ideological. True, ideologues may selectively hark, but for a diehard harker, whether fame reduces or prolongs life expectancy does not matter much. All that matters is the significant correlation between an indicator of fame, for example, people included in some hall of fame, and the age at which people die.

That harking is not ideological may be seen as an advantage, but harking produces junk science nevertheless. Ex post rationalization makes a mockery of the scientific method. Identifying effects that can be argued to be causal and estimating their substantive size is a difficult undertaking. Void of ex ante theoretical guidance, it becomes nearly impossible to distinguish between what is arguably causal and what is merely correlational.

Steve Levitt is one of the most popularly known and famous economists. Not only is he a professor at the University of Chicago, the home university of a large number of Nobel laureates, he also is, together with Steven Dubner, the author of the bestselling books *Freakonomics* and *Superfreakonomics* on questions others never dared to analyze with economic theories and econometric methods. Reading Levitt is fun, and some of his theories and explanations may well be right. But that is not necessarily his prime motivation. According to Levitt, he loves nothing more than "finding the story in data."[11]

"Finding" sounds a lot like harking. But remember that harking is untraceable. When Levitt gets into trouble, it is for other reasons. The most prominent of these is Levitt's arguably most famous correlation, namely, between the rise of abortions in the United States after the *Roe v. Wade* Supreme Court decision in 1973 and the decline in violent crime in the early 1990s, arguing that abortion saved many unwanted children from being born and turning into violent criminals later on. According to Levitt and his co-author John Donohue, this correlation is causal: abortion "appears to account for as much as 50 percent of the recent drop in crime" (Donohue and Levitt 2001: 379). This is a surprisingly bold claim.

As Foote and Goetz (2006) demonstrate, Donohue and Levitt did not include state-specific year trends in their estimations (though they claimed to have done so). If we correct this and change the dependent variable to arrest rates (arrests divided by population), then the findings disappear, becoming substantively trivial and insignificant. How do Donohue and Levitt react to this harsh critique? One would expect that they thank Foote and Goetz for the correction of the error. However, they double down, and in a follow-up publication (Donohue and Levitt 2004) show that they can bring back the results by redefining the abortion measure.

The Rolling Stones were wrong: In empirical research, you can always get what you want. If you only try hard enough.

We do not know whether Donohue and Levitt tweaked their original result. Maybe, maybe not. But they apparently tweaked the results in their response to the Foote and Goetz correction. At the very least they proved our point: that with seemingly small changes to model specification, researchers can tweak their result in the desired direction.

Do Names Kill? The Case of "Female" Hurricanes

One of the rare examples of an empirical analysis that has been openly accused of tweaking is Jung et al.'s (2014) study of the effect hurricane names have on hurricane mortality. Their study claims that hurricanes kill more people if they have feminine names. According to the authors, more people die if a hurricane is named Alexandra instead of Alexander or Victoria instead of Victor.

To quote the authors (Jung et al. 2014: 8782): "Changing a severe hurricane's name from Charley . . . to Eloise . . . could nearly triple its death toll." If you find this implausible, you may be right. When a hurricane threatens to make landfall, both the authorities and the media will warn the public if the hurricanes appear to be or will become severe. The media will communicate the expected strength and severity of tropical storms, their size, expected wind speeds, and where the storm is likely to make landfall days before the storm makes landfall. The name of the hurricane is only one additional piece of information, and people in the hurricane's path would be ill-advised to focus on the hurricane's name rather than its physical characteristics when deciding whether or not to leave the coast.

Still, according to Jung and his co-authors people in the expected path of the hurricane perceive the violence of the approaching hurricane differently when the hurricane has a masculine instead of a feminine name. As is well established by empirical evidence, on average men are more violent than women. Hence, many people will believe, consciously or subconsciously, that hurricanes with masculine names will be more violent and threatening. They are more likely to take precautions when a masculine-named hurricane is approaching, which is perceived to be more violent, than if a feminine-named hurricane is approaching.

Jung et al. (2014) presents some experimental evidence with students and with participants in Amazon Mechanical Turk's online service suggesting that participants in the lab experiment predicted higher intensity for and reported higher perceived risk from masculine- than feminine-named hurricanes and were more likely to intend to evacuate in the face of masculine-named hurricanes. And then they correlate hurricane mortality and the perceived gender of the hurricane name and interpret the correlation as a causal mechanism.

One reason why critics found Jung et al.'s results implausible is because hurricanes are alternately named masculine and feminine, which means that as the sample size grows very large the gender of the hurricane name will be uncorrelated with all relevant aspects of the hurricane. Note, however, that hurricanes are relatively rare events. With only ninety-two hurricanes included in Jung et al.'s empirical study, hurricane mortality may well correlate with hurricane names by chance. Moreover, until 1978 all hurricanes had feminine names; the alternation between feminine and masculine names started only in 1979. During World War II, the US National Weather Service began naming Pacific storms. This measure helped to clearly identify typhoons when more than one storm system existed simultaneously, and it facilitated communication with the navy about potentially dangerous storm systems. Due to its immediate success, the typhoon naming policy was extended to Atlantic storm fronts in 1953 and the hurricane names were communicated to the public. In the same year, the United States abandoned its plan to name hurricanes according to the phonetic alphabet (Able, Baker, Charlie, and so on) largely because of the introduction of a new international phonetic alphabet (Alpha, Bravo, Charlie, and so on). Instead, it began using feminine names for storms so that the first hurricanes were Barbara, Florence, Carol, Edna, Hazel, Connie, Diane, and Ione. Between 1953 and 1978 hurricanes were a relatively rare event. In these years, thirty-one hurricanes hit the United States and all of them had feminine names, a practice that increasingly was considered unfortunate mainly because weather forecasters, who were usually men during those times, developed the practice to talk about these storms as if they were actual women. In 1979 the World Meteorological Organization (WMO) changed the established practice and published lists of low pressure areas that alternated between masculine and feminine names. As a consequence, Bob, David, Frederic, and Allen, the first hurricanes with masculine names, struck the US shore in 1979 and 1980, killing twenty-three people.

A handful of hurricanes have somewhat ambiguous names. The main example is hurricane Sandy, which according to the naming system used by the WMO was a hurricane with a feminine name. Americans use Sandy commonly as a feminine name, but Sandy is also used as masculine name. Sandy was an outlier among hurricanes. It was huge, the largest Atlantic hurricane on record as measured by diameter, and it found its victims in eight countries, ranging from the Caribbean to Canada. Still, Sandy was not the strongest storm on record, making landfall as a mere category 2 hurricane near Atlantic City, New Jersey. Sandy still killed 159 people because it hit areas with populations that had little or no experience with hurricanes.[12]

Looking at the raw data from all ninety-four hurricanes between 1950 and 2012, there can be no doubt that, on average, feminine-named hurricanes have killed more people than masculine-named hurricanes. Figure 5.1 plots all hurricanes in Jung et al.'s dataset against minimum air pressure

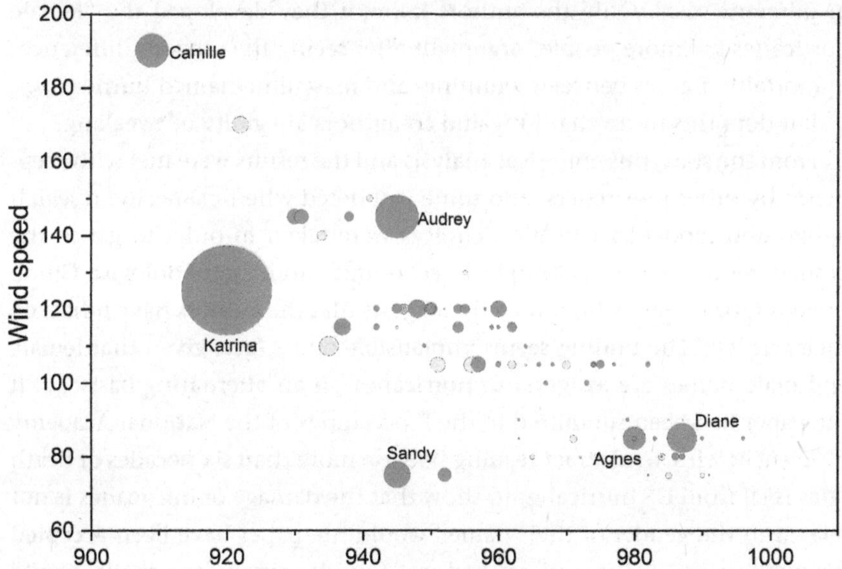

Figure 5.1
Hurricane names, strengths, and mortality.

on the *x*-axis and maximum wind speeds on the *y*-axis, which are two proxy variables for the strength of a hurricane, albeit the lower the minimum air pressure the stronger the storm. The size of each bubble shows the fatalities associated with each feminine-named (dark) or masculine-named (light) hurricane. Clearly, dark dots have, on average, a larger diameter than light dots, meaning feminine-named hurricanes exhibit higher fatalities even if hurricanes show no apparent difference in wind speed and air pressure.

That "female hurricanes" have killed more people than "male hurricanes" is clear from the raw data. But correlation is not causality. As figure 5.1 shows, the statistical association between the gender of hurricane names and mortality relies on a handful of hurricanes with large mortality and feminine names, namely, Katrina, Audrey, Camilla, Agnes, Diane, and Sandy.

Due to the small number of influential data points alone, the tweakability of empirical results is necessarily large. And since Jung et al.'s study was the first empirical analysis of the influence of hurricane names on hurricane deaths, the authors had a "blank slate" for potentially specifying a model that gave them the desired results. The strengths of hurricanes is a necessary control variable. It can be defined, operationalized, and measured in different ways. Only the authors know if they developed the "female hurricanes kill more people" argument after seeing the obvious differences in mortality figures between feminine- and masculine-named hurricanes.

But does this mean that Jung and co-authors are guilty of tweaking?

From the start, this empirical analysis and the results were met with skepticism by other researchers, and some wondered whether specific research design and model specification choices were taken in order to get to the desired result. A good example is economist and methodologist Guido Imbens (2021: 169), who indirectly suggests that the authors have p-hacked their results: "The finding seems implausible on its face, given that female and male names are assigned to hurricanes on an alternating basis. . . . If the paper had been submitted to the Proceedings of the National Academy of Sciences with an abstract reading 'We use more than six decades of death rates [*sic*!] from US hurricanes to show that the damage of hurricanes is not related to the gender of their name,' would the paper have been accepted for publication? If the authors had not found a significant result, would they have moved on to another project?"

Whether Jung and co-authors tweaked or not, some of their sampling and model specification choices look odd and therefore raise suspicion.

One of the major concerns was that the article does not find that feminine-named hurricanes are per se deadlier than masculine-named hurricanes. What they do find is a positive correlation between hurricane names and hurricane mortality if they interact the femininity of the hurricane name with economic damage. They do, however, find the opposite correlation between hurricane names and hurricane mortality if they interact the femininity of the hurricane name with air pressure. Apparently, economic damage and air pressure are both proxies for hurricane strengths, and thus the results should be consistent. But they are not.

Malter (2014) raises a potentially more serious charge. He takes issue with the inclusion of two interaction terms, between femininity and normalized damage as well as between femininity and minimum air pressure, but notes the failure to include a third interaction term between normalized damage and minimum air pressure. If we include this third interaction term, the main effect that Jung and co-authors are after is no longer statistically significant, but with multiple interaction terms the efficiency of each parameter estimate severely declines. Still, methodologists usually advise that one include all interaction terms to prevent a subset of them picking up an effect that belongs to a term that was excluded from the estimations (Brambor et al. 2006).

Other seemingly odd sampling and model specification choices proved, perhaps surprisingly, inconsequential for Jung et al.'s findings, however. For example, some scholars have questioned the legitimacy of using pre-1979 data given that all hurricanes had feminine names up until 1978. The sample consists of 100 percent feminine-named hurricanes between 1950 and 1978 and 50 percent feminine-named hurricanes between 1979 and 2012. Or so you might think. The original analysis gets around this inconvenient lack of variation before 1979 by introducing degree of femininity into the names of hurricanes. They asked nine independent coders to rate historic hurricane names on a scale from 1 (very masculine or male) to 11 (very feminine or female). This creates variation in the femininity scale, although until 1978 all hurricanes had uniformly feminine names. One can question whether coders in the 2010s are qualified to assign a degree of femininity to names starting in 1950 when people's understanding of names was probably different. Yet replacing "degree of femininity" with a binary gender variable does not change the interpretation of results.

The inclusion of pre-1979 hurricanes matters for another reason, too. Jung et al. excluded Katrina (2005), by far the deadliest hurricane, and the second-most deadly hurricane Audrey (1957), from their sample as "outliers," presumably because reviewers could otherwise have thought that these two feminine-named hurricanes alone drive their result. But three of the four next most deadly hurricanes are also from this period: Camille (1969), Diane (1955), and Agnes (1972). This will impact the estimations. Critics argue that this is not because feminine-named hurricanes kill more people but that hurricanes tend to have been deadlier, all other things equal, in the 1950s, 1960s, and the 1970s than during later decades because people got richer and/or lived in more resilient communities; individuals and governments took more precautionary measures; and early warning systems improved in those later decades (Smith 2016). Jung et al. state that initially they included a variable measuring years elapsed since a specific hurricane occurred, which can account for time effects, but dropped the time variable in the main analyses since it proved to be statistically insignificant (Jung et al. 2014: 8785). However, one can restrict the sample to the period 1979 onward and the relevant interaction term remains significant.

Others have demonstrated that the findings are not robust to alternative, equally, or, in their view, more plausible model specifications. Bakkensen and Larson (2014) reveal that the estimates are sensitive to controlling for affected population instead of controlling for population-standardized economic damage.

Simonsohn, Simmons, and Nelson (2020) published a "specification curve analysis" in which they combine a large number of what they regard as reasonable model specification choices and report results on all of them. Of 1,728 specifications that permutate five specification choices with regard to sample and outliers, operationalization of femininity, choice of estimator, conditionality, and control variables, thirty-seven showed a positive association between femininity and mortality consistent with Jung et al. (2014); twenty specifications found an insignificant negative association; and 1,671 model specifications had a positive but insignificant association.

No doubt, the Jung et al. results are sensitive to specification choices. However, the critics seem to have searched for model specifications that alter the results. And if you search long enough, you find the desired differing results. Thus, these critics are potentially guilty of inverse p-hacking: choosing a model specification that gets insignificant results. Clearly, if researchers can tweak results into significance, results can also be tweaked into insignificance. Turning significant estimates into insignificant estimates may be even easier than the opposite. Moreover, all critics focus exclusively on whether the interaction term becomes insignificant rather than what happens to estimated effects. This puts too much analytical weight on p-values and significance.

In the following, we therefore subject Jung et al.'s analysis to a medium-sized number of robustness tests, partly inspired by the detractors of Jung et al. as reviewed above, but we also add some tests of our own. We focus on predicted counterfactual effects, which we define as the change in predicted effects if the actually coded femininity level of hurricane names is counterfactually replaced with the lowest femininity score of 1 for each hurricane, leaving all other variables at values as observed in the sample.

A first series of six tests examine the sensitivity of the estimates to plausible changes in the sample.

1. We limit the sample to the period from 1979 onward.
2. We include hurricanes Katrina and Audrey that Jung et al. excluded on the stated basis that these hurricanes lead to poor model fit due to overdispersion.
3. The third test omits hurricanes Andrew and Diane.
4. The fourth test drops Easy, King and Able: hurricanes that were named before the systematic gendered naming system for hurricanes was introduced in 1953.
5. We remove hurricanes Diane, Camille, Agnes, and Sandy from the sample, the four most deadly hurricanes after Katrina and Audrey.
6. We use wind speed and minimum barometric pressure to match pairs of masculine- and feminine-named hurricanes with comparable propensity scores. Matching gives us a total of thirty masculine- and feminine-named hurricanes each. As table 5.1 shows, in the matched sample we find that on average masculine-named storms cause more economic damage but kill fewer people than feminine-named storms. However, it is difficult to adequately match hurricanes that have made landfall in the United States. The small number of hurricanes with masculine names limits what matching algorithms can achieve.

We also have conducted additional robustness tests on the hurricane name data.[13] Specifically, we replaced the femininity score by a binary gender variable, added the interaction term of economic damage with minimum air pressure, only controlled for the interaction between minimum

Table 5.1
Descriptive Statistics for a Sample of Matched Masculine- and Feminine-Named Hurricanes

	Full sample		Matched sample	
	Feminine	Masculine	Feminine	Masculine
N	64	30	30	30
Wind speed	105.7	102.5	100.8	102.5
Air pressure	963.7	965.3	968.2	965.3
Damage	8353.3	7607.0	6343.3	7607.0
Fatalities	58.2	14.2	39.6	14.2

air pressure and femininity, only interacted wind speed with femininity, replaced the economic damage variable by the hurricane category when it made landfall, controlled for population and population density at the point of landfall (with data taken from Bakkensen and Larson 2014), and varied the functional form assumption underlying the parameter estimate.

What do our robustness tests tell us about the impact of hurricane names on hurricane mortality, and the robustness of these findings?

Hurricanes with feminine names kill more people, no doubt. This still holds if we match the sample in respect to wind speed and air pressure. In the matched sample, hurricanes with feminine names kill on average 58.2 people compared with 39.6 if hurricanes have masculine names. Note, however, that this difference may well be random. One masculine-named hurricane the size of Katrina may well equalize the fatalities of masculine- and feminine-named hurricanes. Impossible? Certainly not, though natural disasters are not independent events: People learn from experience. If this holds true, then another Katrina is less likely to kill the same number of people. If a hurricane the size and force of Katrina were to approach the US coast now, we would expect that more people take precautionary measures.

However, Jung et al.'s inferences are fragile. That an unpredictable property of a hurricane's strength, its economic damage, rather than predictable or known properties such as wind speed, air pressure, or hurricane category is required to get the desired result raises suspicion of tweaking. That the inferences suffer from sensitivity to functional form assumptions is not great for the validity of the inferences either, even if one needs to keep in mind that we do not know the "correct" functional form for the data-generating process.

We do not know whether Jung et al. have tweaked. To us, what smacks most of tweaking is that the significance of their main interaction effect depends on the inclusion of another interaction effect, yet a third interaction effect between the two variables that interact with femininity cannot be included without killing the authors' favorite result. This does not prove tweaking, to be sure, but it does not appear to be theoretically justified.

Thus, the results that link hurricane mortality to the names of hurricane lack robustness. Not only are they sensitive to plausible changes in their model specification. Hurricane mortality estimates are highly tweakable because the distribution of hurricane mortality is highly skewed. Few hurricanes kill many people. Conversely, few people die during the vast majority of hurricanes. A large proportion and perhaps the majority of these deaths appear to be rather random events, driven by what bystanders could call bad luck.[14] Hurricane mortality is largely random and only modestly related to the strength of a hurricane, its category, its minimum barometric pressure, and its wind speed. Mortality is more strongly related to normalized economic damage, but this is not an ex ante measure of strength, and that hurricanes, which cause more economic damage, kill more people is hardly surprising. That does not mean, however, that the economic damage caused by a hurricane kills people. Rather, both factors are correlated because they are influenced by a common third factor, which is not appropriately measured by the variables included in the model.

Most importantly, the results are sensitive to dropping influential cases. Even if we stick to the model specification preferred by Jung et al., when we drop hurricanes Agnes and Diane (1955), the size of the coefficient drops by more than two-thirds and becomes statistically insignificant. This is not because the uncertainty of the data increases, but simply because the predicted slope declines sharply. We are reluctant to make inferences based on a sample that is vulnerable to the exclusion of two extreme cases. If Jung and co-authors tweaked their results, they did a bad job. If they did not tweak, they overgeneralized the results and did not pay enough attention to the serious lack of robustness of their inferences.

Conclusion

Many factors influence the acceptance or rejection of a manuscript by scientific journals, including the relevance of the contribution, the substance of the theoretical argument, the elegance of the writing, and the perceived rigor of the empirical analysis. Increasingly, reviewers and editors will look for the robustness of results and the substantive importance of reported effect sizes. But even the best manuscript has a small probability of getting published in a good journal when the variable of interest is not statistically significant at conventional levels. In other words: A p-value below 0.05 is often a necessary but not a sufficient condition for getting published.

This incentive structure prompts researchers to file away insignificant results into their drawer or alternatively to tweak them into significance. Both strategies inflict damage on the progress of science. With creative sampling decisions or other model specification choices, researchers can tweak the quantity of interest: usually though not necessarily the p-value of the main explanatory variable. Skilled researchers will tweak in ways that are observationally indistinguishable from honest specification choices and will therefore get away with it. Unskilled tweakers may instead resort to hypothesizing after they have seen the empirical results, produced honestly if ineptly. All they need to do is to tell a theory-based story with a prediction that fits the result. This does not mean that most empirical analyses of observational data are tweaked or that inferences based on them are completely wrong. While robustness tests can explore the fragility of findings and thus shed light on their tweakability, they cannot prove actual tweaking. Inferences from observational data analyses can be wrong, whether or not these analyses have been tweaked.

6 Tweaking in Experimental Research Designs

Work hard, squeeze some blood out of this rock, and we'll see you soon.
—Brian Wansink

The Gold Standard

Experiments are the gold standard of empirical research. Or, to be more precise, experiments are *considered* by their proponents to be the gold standard of empirical research (Rubin 2008; Imbens 2010). Experiments have earned this reputation in the natural sciences. Whether this good reputation carries over to the social sciences and life sciences remains contested. When experimental economists Abhijit Banerjee, Esther Duflo, and Michael Kremer won the Sveriges Riksbank Prize in Economic Sciences in 2019 "for their experimental approach to alleviating global poverty," enthusiasts of experimental research designs expected and hoped that this achievement would pave the way for their march toward methodological dominance in the empirical social sciences. It might be too early to tell, but experiments do not yet dominate the social sciences and are not likely to do so in the foreseeable future.

Probably the most important reason why experimental design has not taken over empirical social science research is that experiments cannot answer the vast majority of relevant research questions. Many desirable experiments are not feasible, are too costly, or would be unethical because experiments must not cause substantial harm to participants. As Winship and Morgan (1999: 659–660) point out, "in most social science research done outside of psychology, experimental designs are infeasible. . . . For these reasons, sociologists, economists, and political scientists must rely on what is now known as observational data—data that have been generated by something other than a randomized experiment—typically surveys, censuses, or administrative records." Indeed, most social phenomena cannot be randomized, at least not directly.

Consider these four recent global crises: the COVID-19 pandemic, the Ukraine war, the explosion in energy cost, and the migration crisis. In all these events, experimental designs are usually infeasible. At best, one can study people's perceptions and attitudes toward these events using survey experiments, the poor man's variant of randomized trials.

Nevertheless, even stern critics of social science experiments do not doubt that the randomization of treatments and the control of environmental factors, the two hallmarks of experimental design, have important advantages. In principle. Each feature alone reduces the need to control for confounding factors. In theory at least, they can also function as substitutes: If randomization is perfect and the sample infinitely large, control of environmental factors is not needed. And if control of environmental factors is perfect and the causal effect of interest isolated, randomization of treatment is not necessary. Randomized controlled trials thus kill one bird with two stones. Still, if an experiment is well designed, researchers can be almost certain that any correlation between treatment and outcome has been caused by a causal mechanism linking treatment to outcome. And if, at the same time, the sample is a sufficiently large random draw from the population, the causal effect can be generalized beyond the cases included in the sample.

In most settings, the randomization of treatment is easier than the perfect control of environmental factors, and so it is the former rather than the latter that actual experiments in the social sciences are best known for. However, researchers can often only randomize treatments in a preselected sample. This selection does not matter much when causal effect sizes are homogeneous. If, however, effect sizes vary (as is usually the case in the social sciences and life sciences), the preselection of cases into the sample on which the experiment is run exerts a large influence on empirical results.

The usefulness of experimental designs in the social and life sciences is limited by human nature, at least in contrast to, say, the nature of a particle in a collider. Human beings are inconvenient objects of experimental studies because they not only are heterogeneous but also usually know that they are participating in an experiment. In addition, they learn from each other and react to externalities caused by the behavior of others. The behavior of humans is, almost by definition, not independent of the behavior of others. Human beings are social.

This apparently violates the stable unit treatment value assumption (SUTVA). Although SUTVA does not sound like a big thing, the assumption implies that experimental research is only unbiased if the potential outcome for one unit is independent of the assignment of treatment status to other units. This assumption of stable unit treatment seems to be a more innocuous assumption in medical research where the effect of a treatment on one patient is usually independent of the effect of treatment on other patients. The effect of a painkilling pill for one person is independent of the painkiller's effect for another person. In many social science experiments, however, the assumption is more problematic and often violated.

Experimental design relies heavily on the reputation of experiments in the hard sciences, especially physics. However, the underlying data-generating processes have little in common, and thus the properties of social and life science experiments differ significantly from those of physics experiments. While experiments may define the gold standard of research methodology, not every experiment meets these standards.

Why Results from Experimental Designs Can Be Tweaked

Virtually all articles retracted for data fabrication and manipulation are experimental in design. As such, this proves nothing. Any overrepresentation of experimental research designs among fraudulent articles could be due to greater scrutiny of experimental designs. However, there are structural reasons why experimental designs are vulnerable to fraud. At first glance, this seems surprising. In one sense, experimental results are less open to tweaking than observational data analyses because experimental designs are less dependent on estimation models that account for confounding factors. And yet, in important other ways, experiments take matters out of the frying pan just to throw them into the fire. Experimental designs give researchers an almost complete control over the design of the experiment. This becomes clearer with another label used for these designs, namely, randomized *controlled* trials. And with control comes tweakability.

Experiments provide paradisical conditions for fraudsters. Researchers have almost complete control over the design of the experiment, and if they wish, they not only conduct the experiment but also collect all the data. This means that they can manipulate the data in many ways without fear of detection. Even if they publish the final estimation data, as long as they keep the raw data private, manipulations cannot be proven. It is thus probably not by chance that the social science discipline that relies the most on experiments, social psychology, appears to suffer most from data fraud and retractions.

If fraudsters turn a "2 into a 4 and a 3 into a 5," as Diederik Stapel admitted to doing, who is to say that the 5 originally was a 3 and the 4 a 2? This is more difficult in observational data analyses where the data are often not privately collected. Hence, we expect to find more tweaking in observational data analyses and more data manipulation and fabrication in experimental research designs. It is therefore not surprising that the majority of detected outright data fraud comes from experimental designs. But experimental researchers do not need to rely on data manipulation. Tweaking is still an option for experimental researchers as well, and since the majority of experimental researchers are likely to shy away from outright fraud, they can resort to tweaks instead. Since tweaking is seen as a "lesser sin," "not really cheating," and a "cavalier act," the psychological cost of tweaking is lower.

Experimental designs offer fewer opportunities for tweaking than observational data analyses. Since randomization of treatment status asymptotically eliminates the influence of confounders on outcomes, analyses of experimental data usually rely on parsimonious empirical models, calculating the mean difference between the treatment and the control group. This has implications for tweakers: The lower model dependence of experimental designs reduces model uncertainty and the number of necessary specification choices. As a consequence, fewer tweaks are available to researchers seeking to influence quantities of interest in experimental research. But fewer does not mean none. And there always is the option to manipulate data points.

There are three main tweaking strategies in experimental designs: First, the experiment can be designed to bias the result toward the desired outcome, providing evidence for the hypothesis being tested. Second, experimental designs invite fishing strategies. Researchers can always use more than one treatment and observe more than one outcome in experimental studies. And some combination of treatment and outcome choices will probably yield a publishable result. Third, few effects are unconditional. Instead, they vary between identifiable subgroups of the population. This allows tweakers to manipulate the sample to get to the subgroup in which they find the empirical result they want or to drop inconvenient cases from their sample that prevent them from finding the desired result.

Before we explore these three main tweaking strategies, it is important to note that fraudsters may be incompetent, make plenty of honest mistakes, tweak results, and engage in outright fabrication and manipulation, all at the same time. We know of no better example for the mélange of incompetence, manipulation, and tweaking than Brian Wansink. Wansink's career began as an undergraduate student at Wayne State College, Nebraska, followed by an MA in journalism and mass communication from Drake University in 1984. Six years later he earned a PhD in marketing from the Stanford Graduate School of Business. His first academic positions were held at the Tuck School of Business at Dartmouth College (1990–1994), from where he stumbled up the ladder to Wharton Graduate School of Business at the University of Pennsylvania (1995–1997) and then on to the University of Illinois at Urbana-Champaign (1997–2005) where he became professor of, well, apparently everything ("Marketing, Nutritional Science, Advertising, and Agricultural Economics"). He finally moved to Cornell University's College of Agriculture and Life Sciences where he was also director of the Food and Brand Lab. So far, so good.

Brian Wansink's academic career ended with a big bang when he resigned from his position at Cornell after the university admitted that he had committed extensive scientific misconduct. In December 2018 the university ruled that Wansink had engaged in research practices that "included data falsification, a failure to assure data accuracy and integrity, inappropriate attribution of authorship of research publications, inappropriate research methods, failure to obtain necessary research approvals, and dual publication or submission of research findings" (Lee 2018).[1]

Wansink's downfall was triggered by an investigation into a series of articles based on analyses conducted by his graduate students at Cornell. The first set of analyses Wansink conducted produced null results. Therefore, Wansink asked his students to turn these results into something *significant* and publishable. In January 2017 Wansink even published a blog post in which he openly told the public how he had asked a grad student to "salvage" conclusions from a study that had null results. According to Retraction Watch, Wansing wrote: "When [the grad student] arrived, I gave her a data set of a self-funded, failed study which had null results. . . . I said, 'This cost us a lot of time and our own money to collect. There's got to be something here we can salvage because it's a cool (rich & unique) data set.' I had three ideas for potential Plan B, C, & D directions (since Plan A had failed)."[2] This exercise resulted in four articles that were later called the "pizza papers" (because Wansink asked students in a pizza restaurant to "salvage" seemingly unpublishable results), by a group of methodologists who reanalyzed them (van der Zee et al. 2017). Tim van der Zee and his co-authors concluded that they did not remember "encountering a set of articles with as many inconsistencies and unresolved questions" (van der Zee et al. 2017: 7). They also said that they "made generous allowances for rounding, considered all possibilities for test statistics, . . . made multiple passes within and across the articles to try and identify the correct sample sizes, . . . checked granularity errors with an online web application, with an Excel sheet, with R code, and by hand" (9).

The Wansink case demonstrates that publications can result from any combination of outright fraud, tweaks, and sheer incompetence. While Andrew Gelman is otherwise scathingly critical of Wansink and regards him as a pure careerist fraudster, he admits that he's probably not a bad person, though he's a terrible scientist: "Based on his blog post and his responses to comments, he seems genuinely interested in making people happy. The quality of the work doesn't matter at all."[3] The incompetent but well-meaning Wansink fabricated, manipulated, tweaked, and made mistakes. All at the same time. Just to make the world a happier place.

Biased by Design

The researchers behind the famous RAND Corporation's Health Insurance Experiment may not have consciously tweaked their experimental design to influence results, but this example demonstrates how an experimental design can bias results in favor of a particular outcome. The RAND experiment was a large-scale and expensive field experiment funded by the US Department of Health, Education and Welfare that aimed at investigating whether generous health insurance leads to costly "inappropriate or unnecessary" health interventions. The design of the experiment was simple: Health insurance schemes with varying private co-insurance rates (out-of-pocket contributions) were randomized among 5,809 voluntary participants. The co-insurance rates varied between 0 percent, 25 percent, 50 percent, and 95 percent. Out-of-pocket contributions were capped at $1,000 per year.

The experiment showed that participants in more generous plans consumed more medical treatments and procedures than participants in plans with higher out-of-pocket contributions. Those who benefited from free care received about 50 percent more medical interventions, resulting in approximately 33 percent higher costs on average. According to the researchers conducting the experiment, cost-sharing reduced "appropriate or needed" medical care as well as "inappropriate or unnecessary" medical care. Since no difference in health status was observed, the case for relatively high private co-insurance rates was strong. Requiring individuals to pay a higher share of the medical costs via out-of-pocket contributions reduces "inappropriate or unnecessary" medical care. While some demand for "appropriate or needed" care might be discouraged, the dampening effect on this socially desirable spending was not large enough to have a noticeable negative impact on health outcomes.

The RAND experiment was biased by design: a bias that exaggerated the evidence for its main findings. The basic finding is plausible and follows from the "tragedy of the commons": Public goods such as free health insurance are being overused. But the extent to which generous health insurance led to unnecessary health care was at least partly caused by the incentives built into the experiment's design. The experiment took place in the shadow of the real-world health insurance plans to which participants returned at the end of the experiment. Participants knew that the experiment was not going to last forever and that their experimental coverage was going to end eventually. If they had a more generous health insurance in the experiment than they had in real life, they faced incentives to move medical interventions forward in time. In contrast, participants who were assigned to a less generous insurance scheme than they had outside the experiment were incentivized to delay interventions. This exemplifies the trade-off between the advantages and disadvantages of field experiments versus lab experiments. More realistic treatments can be studied in field experiments, but better control of environmental factors and mitigation of strategic participants' behavior in response to the treatment can be achieved in lab experiments.

The RAND experiment created incentives for strategic behavior that explains the higher health spending of participants with generous insurance during the experiment. The experiment incentivized behavior that biased this result upward. Was this an honest mistake, an inevitable problem of an experimental study, or a deliberate manipulation of results? We do not know. In the 1980s, the US government increased private co-insurance schemes. The RAND experiment provided important arguments for those who pushed for this policy change (Nyman 2007). This is more problematic because strategic behavior by participants was not the only aspect of the experiment that biased the results. In a review article written three decades later, Aron-Dine, Einav, and Finkelstein (2013: 2020) conclude that "despite the potential for substantial bias in the original estimates stemming from systematically differential participation and reporting across experimental arms, one of the central contributions of the RAND experiment is robust: the rejection of the null hypothesis that health spending does not respond to the out-of-pocket price. Naturally, however, these potential biases introduce uncertainty about the magnitude of the impact of the different insurance plans on medical spending."

Unlike field experiments, in which treatments are randomized in real-world settings, laboratory experiments allow one to hold environmental factors constant, but they usually drive a large wedge between the treatment in the artificial lab setting and the real-world phenomenon about which researchers are trying to make inferences. They allow one to study not actual behavioral responses but only perceptions, opinions, and attitudes. How these weak factors translate into behavior in the real world remains unclear, but they certainly do not translate one-to-one into behavior. This ambiguity can be exploited by tweakers who frame the treatment in ways conducive to finding the desired results and then make more sweeping claims about what their results imply about the real world.

Lab and survey experimental results deliver potentially biased results because participants are usually aware that they are contributing to an experiment, being observed, monitored, and controlled, and they adjust their behavior in the experiment accordingly. They participate voluntarily in the experiment, which limits the generalizability of the experimental results to a broader population. Even in field experiments, participants may realize that they are participating in an experiment. And then they feel observed, which leads to more pro-social behavior or the expression of more pro-social opinions, especially if the experimental design does not avoid trigger terms such as altruism or cooperation (Winking and Minzer 2013).

If researchers want to demonstrate the existence of altruistic behavior, it is possible to bias the results in the desired direction by letting participants know that they are being observed. Those conducting the experiment can consciously influence participants' behavior in a particular direction by the way they set up the experiment or by the way they communicate with the participants. Of course, they shouldn't. But then, tweakers shouldn't tweak.

Fishing

Experimental results can be manipulated via fishing expeditions, the second main tweaking strategy. Experiments often collect more information, use more than one treatment, and observe more than one outcome. This seems to be an efficient solution to the problem of scarce resources: the experiment does not have to be repeated every time for every combination of treatment and outcome. However, it clearly allows fishing for significant associations between treatment and outcome. If you collect data on ten different treatments and ten different outcomes, some associations are bound to be significant.

Fishing is not limited to collecting data on multiple treatments and multiple outcomes. Experimental researchers can fish for publishable results by repeating the same experiment over and over again, until the data finally confess to what the researcher believes they should be saying. Even in cancer research, one of the most sensitive research areas with numerous lives at stake, not all published results can be replicated. One of the more bizarre cases of failed replication concerns an article on tumor growth in animals published in *Cancer Cell* (Couzin-Frankel 2013). In this case, the scientist who failed to replicate the published results, Glenn Begley, invited the author of the original paper (whom Begley declined to name) to a breakfast during a scientific conference that both scientists were attending. As the conversation progressed, Begley revealed that he had tried and failed to replicate the analyses reported in *Cancer Cell*. He was stunned when the author of the original article admitted that he himself had failed in eleven out of twelve trials he conducted but had published the one trial that showed significant results. According to Couzin-Frankel (2013: 68), "Begley was aghast. 'I thought I'd completely misheard him,' he says, thinking back on the encounter. 'It was disbelief, just disbelief.'" This seems to be an extreme case of fishing: Run and run and run the experiment until you accidentally find a significant result in one trial and, voilà, you can submit your work to a journal. Apparently, some established researchers seem to have little awareness of where scientific integrity ends and fraud begins.

Another tweak within the fishing strategy is known as "optional stopping" and "data peeking." These strategies work by starting with a small number of cases in the sample, running the experiment, checking to see whether a significant result was obtained, and, if not, successively adding more and more cases to the experiment until results become publishable. Larger sample sizes improve the odds of finding significant results. Since, all other things equal, the p-value decreases as the number of observations increases, tweakers with deep pockets can add cases until the standard error of the estimate becomes sufficiently small and the p-value significant.[4]

A weird example of an experimental study whose author has been accused of fishing is Daryl J. Bem's (2011) article "Feeling the Future: Experimental Evidence for Anomalous Retroactive Influences on Cognition and Affect." What makes this case bizarre is that the causal mechanism required for the alleged effect contradicts everything we know about physics, including the mysterious quantum entanglement. Bem seems to be aware of the tensions between his results and mainstream physics but suggests that future advances in physics will demonstrate how and why the effects he claims to have found can be and are real.

Bem (2011) argues that information does not always travel forward in time, but instead occasionally travels backward. He claims to have found strong evidence from nine experiments for what he calls precognition and premonition of a future unknown event. Note that we are not talking about what economists and psychologists refer to as "rational expectations," where individuals form current expectations about something they suspect to happen in the future based on some information they have in the presence. Instead, Bem argues that agents sense the future. In his experiment, "information travels backwards in time" (Bem 2011: 423). Or so Bem claims.

Science is well advised not to rule out findings on the basis that they are deemed impossible according to current scientific understanding or intuitively implausible. Quantum entanglement describes the relation between a group of particles in a way such that the quantum state of each particle of the group cannot be described independently of the state of the others even if these particles are separated by an arbitrarily large distance. Intuitively plausible? Certainly not. Not even to Einstein.

We do not argue that Bem's findings are definitely impossible. At least not less possible than many of the results in quantum physics. But they are so implausible that Bem's research set off alarm bells as soon as his article was published in the highly ranked *Journal of Personality and Social Psychology*. Stuart Ritchie, a PhD student back then, together with two other psychologists replicated Bem's experiment but, surprise, surprise, failed to arrive at Bem's results, which sparked Ritchie's interest in exposing scientific fraud as an important part of his career as a science journalist (Ritchie 2021: 3). Wagenmakers et al. (2011) and Rouder and Morey (2011) soon questioned whether the one-sided tests for statistical significance applied by Bem (2011) were correct and suggested they were biased toward finding significance. Alcock (2011) questioned the integrity of Bem's experiments, which he finds insufficiently described to allow replication, and pointed out that Bem changed the procedures while the experiment was underway. Francis (2012) accused Bem of fraudulently reporting only selected supportive results. He based this charge on his assessment of how likely it was to find supportive evidence in nine out of ten described investigations (one of the nine experiments contained two investigations). Ulrich Schimmack and three of his colleagues go furthest in alleging fraud in their letter to the editor of the journal and requesting the retraction of Bem's article. They suggest that Bem has most likely manipulated his findings through a fishing strategy. Having had access to the original data, they found that the effects tended to be larger with small samples and smaller with large samples. From this observation they concluded: "The most plausible explanation for this pattern of results is that Bem aborted studies when the results were not promising and continued data collection as long as the data trended in the right direction."[5]

Did Bem tweak? Based on the available information only Bem could know. Alcock (2011) finds it suspicious that Bem had written advice for researchers on how they can best hypothesize after results are known. It is worth quoting Bem at length here: "Examine them from every angle. Analyze the sexes separately. Make up new composite indexes. If a datum suggests a new hypothesis, try to find additional evidence for it elsewhere in the data. If you see dim traces of interesting patterns, try to reorganize the data to bring them into bolder relief. If there are participants you don't like, or trials, observers, or interviewers who gave you anomalous results, drop them (temporarily). Go on a fishing expedition for something, anything—interesting. No, this is not immoral. . . . In the confining context of an empirical study, there is only one strategy for discovery: exploring the data. . . . Think of your dataset as a jewel. Your task is to cut and polish it, to select the facets to highlight, and to craft the best setting for it. Many experienced authors write the results section first."[6] This sounds a lot like someone who will not shy away from tweaking experimental results.

The ability to go on a fishing expedition can readily be incorporated into experimental designs. The easiest strategy is to use different treatments and observe different outcomes. The law of large numbers suggests that if we combine a large number of treatments with a large number of outcomes, some combinations will yield significant results of sufficient magnitude.

Sampling and Resampling

The fact that inferences based on experimental designs *can* be perfect does not mean that they actually *are* perfect. In real experimental research the size of the actual sample is often small, and so the impact of confounding factors has at best been reduced but not completely eliminated. Sample size issues are compounded if other factors condition the strength of the treatment effect. These conditioning factors must be randomized in the experiment if researchers are to make valid and reliable predictions about outcomes. Since the number of participants would need to be large enough to randomize treatment status for each of the various relevant values of the conditioning factor, the required sample size increases exponentially. Experimental studies that model conditioning factors are therefore extremely rare.

Sample size issues affect all experimental designs. Other caveats affect different types of social science experiments to varying degrees. Field experiments tend to have better concept validity than lab experiments. The treatment in field experiments closely resembles the real-world treatment, about which researchers make inferences. This is great. Field experiments also stand a greater chance of studying a more representative sample. Not so great is the context dependency of results and the limited ability to hold environmental factors constant. All social science experiments can only identify the "true causal effect" for the sample at the point in time and within the context the experiment was conducted. If causal effects vary across space and time and context—and this seems highly likely for all "treatments" in the social sciences—even "true" effects cannot be generalized. Manski (2013: 36) points out that the proponents of experimental designs "prefer achieving credible certitude about an easy-to-study population to achieving credible partial knowledge about the population of policy interest."

In lab experiments, only readily available volunteers are recruited, since individuals cannot be forced to participate in an experiment. The self-selection of participants always raises doubt about whether any effect found can be generalized to the population of interest. Lab experiments allow researchers to randomize treatment across participants, but they usually cannot randomly draw participants from the population. Many university labs operate participant pools: students who regularly participate in their experiments. These students do not represent the average person on the street. They are usually younger, healthier, and smarter (or at least think they are) than the average person. In a meta-study analyzing thirteen studies on whether students behave differently from professionals, Fréchette (2011: 33) concludes that "professionals and students have different behavior. . . . Students are closer to the theoretical prediction than professionals."

Sampling decisions exert a strong influence on experimental empirical results because in the social and life sciences, context and environmental factors do matter. Causal heterogeneity, which is at the heart of the limited transferability of experimental results, can arise from individual characteristics of participants: People vary in their genetic makeup, their moods, their interests and preferences, their resources, their skills, their opinions, their knowledge, and other characteristics. But causal heterogeneity may also result from the institutional, societal, and macroeconomic setting in which the experiment takes place (Pritchett and Sandefur 2015). Consequently, an estimate derived in an urban setting in a developed country in 2019 cannot necessarily be generalized to different settings or different times or to a seemingly similar setting at the same time. As Jim Manzi, a strong advocate for social science experiments, admits (2012: xiv): "We can't conclude that if literacy program X works in Norfolk, then it will work everywhere. The real predictive rule is usually closer to something like Literacy program X is effective for children in urban areas, and who have the following range of incomes and prior test scores, when the following alternatives are not available in the school district, and the teachers have the following qualifications, and overall economic conditions in the district are within the following range. And even this predictive rule stops working ten years from now, when different background conditions obtain in the society." He contrasts this to medical trials: "We can run a clinical trial in Norfolk, Virginia, and conclude with tolerable reliability that 'Vaccine X prevents disease Y.'"

Tweakers can find a selection criterion that restricts the sample in a way that produces the desired results in some subgroup of the full sample. If the treatment effect remains insignificant in the population, perhaps it may still become significant for women only, or for men only, or for people younger than sixty or older than twenty-five? If both the treatment and control groups are large enough, researchers can stratify in more than one way. Chances are that the treatment effect will be what the scientist prefers in at least one of the many possible subsamples: significant, substantively large, fit for publication.

One of the most striking examples illustrating this tweaking strategy involves Scott Harkonen, the former CEO of InterMune. The company organized a randomized trial for a drug designed to improve the survival rate for patients with a fatal lung disease called idiopathic pulmonary fibrosis. The company saw a $2 billion market opportunity: In the United States alone, approximately 50,000 people are diagnosed with the disease, which slowly fills the lungs with scar tissue and on average kills patients within two or three years. The study compared the survival rates for all treated and control patients and found a positive effect: "Disease progression or death occurred in 46 percent of those on the drug and 52 percent of those on placebo."[7] However, with a p-value of 0.08, this small effect did not reach conventional levels of statistical significance. The company set out to slice and dice the sample to find a statistically significant effect in a subgroup. InterMune found that for a specific subsample of participants with mild to moderate variants of the disease, the drug was associated with an improved survival rate: "Only 5 percent of those taking the drug died, compared with 16 percent of those on placebo."[8] And after dropping all cases of severe disease, the p-value was highly significant, at 0.004. As the company stated in a press release: "The drug reduces mortality by 70 percent in patients with mild to moderate disease."[9] The company's CEO, Scott Harkonen, was later sued and convicted for issuing a misleading press release that exaggerated the results and provided insufficient context. To make matters worse, the company failed to replicate the result for the subgroup of people with mild to moderate variants of the disease. In a follow-up study with 826 patients with mild and moderate symptoms at eighty-one hospitals, 15 percent of the treated patients had died, while the death rate in the control group on placebo was only 13 percent.[10] Not a positive effect, and not statistically significant either.

A treatment may well work only in identifiable subgroups. As Manski (2013: 63) explains: "While it rarely is credible to assume that individuals respond identically to treatment, it sometimes is credible to assume that treatment groups have identical response distributions." However, if researchers restrict the experimental samples to different subgroups to find what works, these results need to be tested in a new experimental setting with a new sample. Ideally, one has a theoretical explanation for the causal mechanism of why the treatment works in one group but not another. Experiments alone are of limited use in this regard, as they provide only limited insight into causal mechanisms (Imai et al. 2013).

An alternative tweaking strategy to restricting the analysis to a favorable subset of the sample is the elimination of selected cases that propel estimates in the "wrong" direction. Assume a treatment in a medical trial reduces fever of male patients by 1.0 degrees Celsius on average but reduces the fever of females by only 0.2 degree Celsius. If an equal number of men and women participate in the experiment, the local average treatment effect is 0.6. It may or may not be statistically significant depending on the sample size and the variability within the group of men and the group of women. But for the sake of argument assume that the effect is significant within the group of men, but neither significant in the group of women nor in the full sample. One tweak is to report the result for the subgroup of men only, similar to what InterMune did in its trial. However, this tweak is not commercially optimal because the treatment is reported to be effective only for men.

There exists a commercially more attractive tweaking option available. Tweakers who know about the outcome can find reasons to exclude some women from the sample. In experiments, one can usually find a justification for dropping some cases on the basis that they did not comply with their assigned treatment status or other parts of the experimental protocol or that researchers discovered after their enrollment that they had some existing precondition that justifies excluding them as outliers. In our hypothetical medical trial, if we found a selection rule that increased the proportion of men to 80 percent of the participants, the local average treatment effect would increase from 0.60 to 0.84 and become statistically significant. The effect size is not as large as if the treatment were alleged to be effective only for men, but because now it can be sold to the other half of the population too, this tweak more than makes up for it.

If the selection of cases into the sample on which the experiment is run exerts a strong influence on empirical results, researchers can directly influence results by adding or dropping selected cases from the experiment. Indeed, combine a certain arbitrariness in selecting participants with the fact that a surprisingly large share of experimental researchers collect their own data, and one has near-perfect control over results from experimental designs.

Are Men More Altruistic Than Women?

We use unpublished data from repeated lab experiments of a dictator game with 10,320 university students in four countries to show how experimental results can be tweaked by resampling and other model specification choices.[11] Dictator game lab experiments are not randomized trials. They do not randomize treatment and no control group exists. However, dictator games control the setting in which participants interact, and they usually hold the treatment constant.

In the simplest form of the dictator game, participants are given an amount of real or hypothetical money as windfall gain. They are then asked how much of that money they are willing to give to other participants. As more money is always better than less money, rational and purely selfish individuals keep 100 percent of the money. Accordingly, dictator game results "measure" the degree of pure altruism by the share of money individuals choose to give to others, since keeping 100 percent of the windfall profit does not come with an obvious cost.

In theory.

In practice, however, the behavior of participants is observed and recorded. It is a lab experiment after all. And the participants know they are being observed. Thus, the experiment tests the extent to which participants are willing to share when they feel observed and when their social reputation is at stake. Most participants have been educated and socialized by their parents and teachers to believe that selfishness is inferior to pro-social orientation and cooperation. As a consequence, keeping 100 percent of the money does come with non-monetary costs.

How much are participants willing to share in order to be one of the good guys? According to Levitt and List (2007), more than 60 percent of participants share money and transfer on average approximately 20 percent of their windfall gain. Evidence suggests that women give away more than men (Eckel and Grossman 2008). Whether women are intrinsically more prone to altruistic behavior, whether they respond more strongly to feeling observed, or whether they are, on average, socialized in a somewhat different way than men remains unclear but is not of interest here.

Given the lack of randomized assignment to treatment status—the gender of participants is a given for the experimenter—those reviewing a manuscript based on the findings from dictator game experiments are likely to expect researchers to control for some observable characteristics of participants. In our dataset there are two control variables accounting for the mathematical skills of participants and their degree of risk aversion. Ideally, one would want to control for more participant characteristics: their study major, socioeconomic status, and age, to mention a few obvious potential confounding factors. The small number of control variables in our dataset could be an issue if we were substantively interested in the effect of sex on altruistic behavior. However, we are interested in the ease with which experimental data with sufficient heterogeneity among participants can be tweaked.

Table 6.1 provides summary descriptive variable information on the student participants in each of the four countries. Almost 45 percent of the 10,320 participants come from the United Kingdom, followed by Russia (25%), and Chile (22.5%), with the small remainder coming from China, which is the only country in which men and women are not approximately equally represented (female students make up almost 60 percent of participants in China). On average, participants parted with 295 of their 1,000 currency equivalents. Women were twenty-four currency units more generous, and men twenty-four units less generous, than the average person.

Results vary widely: Chilean female students were particularly generous compared with men, giving away nearly sixty-four currency equivalents more than male students, while Chinese female students gave away only twenty currency equivalents more than their male counterparts on average. Altruism also varied across the participating universities in countries, being highest in China and lowest in the United Kingdom. Perhaps communist ideologies have a lingering effect, or maybe this is just how Oxford University selects students.

Table 6.2 shows the gender gap in altruism. In a simple set of analyses regressing individual sharing behavior on gender, we find that, without

Table 6.1
Descriptive Statistics and Average Offer by Country

	Average offer	Numerical skill	Average risk aversion	Number of participants
Men and women	295	10.6	5.59	10,320
Chile	378	8.6	5.94	2,320
China	408	7.5	6.04	840
Russia	263	11.2	5.10	2,560
United Kingdom	251	11.8	5.61	4,600
Men only	271	11.4	5.50	5,250
Chile	348	9.2	5.91	1,210
China	396	8.4	6.38	340
Russia	239	11.9	4.89	1,330
United Kingdom	232	12.7	5.49	2,370
Women only	319	9.8	5.70	5,070
Chile	412	7.9	5.98	1,100
China	416	6.9	5.80	500
Russia	288	10.5	5.33	1,,230
United Kingdom	270	10.9	5.74	2230

Table 6.2
Gender Gap in Altruism

	Chile	Chile	China	China	Russia	Russia	UK	UK
Women	63.9	70.2	20.2	18.2	48.9	23.3	37.8	17.9
s.e.	(29.1)	(30.3)	(25.2)	(28.4)	(29.4)	(29.9)	(17.7)	(18.9)
p-value	0.046	0.036	0.481	0.566	0.115	0.446	0.043	0.354
Controls	No	Yes	No	Yes	No	Yes	No	Yes

controls, we get significantly negative estimates of men's generosity for both Chile and the UK, but not for China and Russia. Once we control for numerical ability and risk aversion, only Chile shows a significant gender gap in altruism.

Participants in Chile are different from participants in these other countries, although it remains unclear whether these effects vary because participants, universities, cities, or cultures differ. We are interested in learning more about the contexts that matter, and why they matter and affect men and women differently. None of these questions are answered by these dictator game experiments.

Researchers seeking evidence to support the hypothesis that women are more generous than men in dictator games will not be satisfied with the above results, even if all point estimates indicate less altruism among male participants. If researchers (correctly) anticipate that they will not get an insignificant result published, they will dislike that the global sample estimate is only significant without control variables included (not shown in table).

But this is not the end of the research project. Tweakers could continue to add more experimental sessions, especially in Chile. With a few more sessions from Chile added to the full sample, the p-value might give up its stubborn resistance to fall below the magical 0.05 level. But that requires more money (unless our tweaker decides to just duplicate a few "nice" observations from Chile). Alternatively, they could just publish the result based on the Chilean subsample. It is quite common for lab experiments to be based on participants from only one country. Yes, this obviously reduces the external validity of the results. However, external validity had been absent from the beginning. The participants are students, and they are too young, too educated, and perhaps too pro-social to be representative of the broader population. Participants' behavior is observed, which means that we cannot generalize results to situations in which behavior is not observed. Perhaps we can generalize results from the Oxford University sample to a Cambridge University sample. But certainly not to the UK population. And we are not sure about Cambridge either.

There are other options for potential tweakers. One is to include country-fixed effects in the full sample estimation. Men are now estimated to give 30.4 fewer currency equivalents on average than women, and the estimate comes with a delightfully low p-value of 0.026. The downside is that our tweaker needs to justify the inclusion of country-fixed effects. This is of course possible but ultimately may cast doubt on whether these experiments identify the true causal effect.

Is it a promising tweaking option to keep peeking at the results as additional experimental sessions come in?

We find that with the data we use, the population size has little effect on results unless we exclude Russia from the sample. Whether this can be justified is a different question. It may be, however, that economics students are overrepresented in Moscow and that this suffices as justification for dropping students from Moscow. Once we exclude Russia and analyze data from eight sessions in each remaining country, the desired results appear: Men give 42.7 currency units less than women, with a p-value of 0.041.

One can, however, also tweak the sample to obtain the opposite result. That women are on average more generous than men is only mildly interesting, as others have shown the result before. The opposite effect, that men are under identifiable circumstances more altruistic than women, would certainly be more exciting for psychologists. And where there is a will, there is a way. If we restrict the sample to the 6,440 most generous participants, those who give away more than 25 percent of their currency units, men are on average more generous than women. In this subset of participants, men pass on 15.6 currency units more than women. And with a p-value of 0.031 the result is significant at conventional levels. Bingo.

Conclusion

Experimental designs are widely regarded as the gold standard of empirical inquiry. However, what works well in the natural sciences does not neatly transfer into the social and life sciences. These scientists are ill-advised to generalize properties of experimental designs from quasi-deterministic causal mechanisms with homogeneous effects, as in the natural sciences. In the social sciences, they are confronted with probabilistic causal mechanisms and heterogeneous causal effects across socially interacting human beings.

The limitations of social science experimental designs are one thing; their vulnerability to data fraud another. Experimental research can be manipulated and fabricated with great ease since researchers often conduct their own experiments, based on their own experimental designs; collect their own data; and transform the raw data they have collected to the data they eventually analyze. Given that experimental data are often private, we should not be surprised that the vast majority of detected data fraud comes from experimental studies.

Experimental researchers can manipulate their data until they get the results they want, but they also have less radical and less noticeable options at their disposal. The options for tweaking experimental results are manifold: Researchers might introduce biases in the design of the experiment or artfully tailor the composition of the selected sample. They determine the sample of participants and may observe multiple outcomes and treatments, which allows them to fish for results in their own private data pond.

The vulnerability of experimental designs to tweaking and other scientific fraud has triggered preregistration of the experimental design and the analysis plan as a response. However, preregistration opens the floodgates to an infinite regress. Researchers always have the option to collect information before registering an experimental design and this information can be used to bias results in a particular direction. Sure, conducting a pre-experiment before registering the experimental design is costly, but many experiments are inexpensive and from the perspective of a fraudster, money spent on a pre-experimental study appears to be money well spent. Not all that looks like gold is actually Au79. Gold standard perhaps, but the gold produced by experimental studies may occasionally result from alchemy. And if you scratch at the surface, something else, iron or lead, becomes visible.

7 Tweaking in Causal Models

P-hacking is not related to researcher pedigree.
—Abel Brodeur, Nikolai Cook, and Anthony Heyes

The True Causal Effect

Identifying the true causal effect of a treatment on an outcome should be dead easy: Just treat and not treat the same person at the same time, observe the two outcomes, and calculate the difference. This difference between the observed and counterfactual outcomes is the treatment effect for the treated person at the time of treatment. If we do the same analysis for the population, we get the true average treatment effect for the population, at least for that one moment in time. And if we repeat this exercise often enough, we can see if and how the average treatment effect of the population changes.

Except there is one major flaw: This research design is not feasible. We cannot treat and not treat the same person at the same time, an impossibility known as the fundamental problem of causal inference (Holland 1986). Fortunately, over the past fifty years or so, clever research designs have been invented to mitigate or circumvent the fundamental problem of causal inference. The randomization of treatment in experimental designs is one of these strategies that work well where they are feasible. But for many, if not most, research questions, they are not feasible.

For situations where experiments are infeasible or undesirable, methodologists have developed "causal models" that seek to identify causal effects in observational rather than experimental data by trying to bring some element of randomization into observational data analyses. Fundamentally, causal models seek to find or restrict samples to settings where the treatment and control groups are as similar as possible on all relevant factors other than treatment so that the assignment of treatment status can be argued to be "as if" random rather than truly randomized. If that works, these research designs can be used with all types of observational data and at the same time promise to reduce the dependence of traditional models on model specifications. You can't have the cake and eat it? Not if the promises of the proponents of causal models come true.

According to its developers and users, causal models are at the forefront of a methodological revolution sweeping through the social sciences. Causal models are as close as possible to experimental designs—so close that their fans call them quasi-experimental: "Quasi-experimental designs are broadly characterized by the assumptions under which the (non-random) treatment assignment mechanism is as good as randomized" (Sekhon and Titiunik 2017: 3).

Causal models neither randomize treatments nor hold constant the environment in which treatment occurs, as experiments do. These models therefore do not really share the superior properties of experiments. They are, instead, observational studies with a twist. These twists vary across the identification strategies employed by different causal models. Causal models provide ample opportunities for tweakers. The best strategy to tweak causal models is by manipulating the case selection algorithm. Since causal effects often tend to be heterogeneous, case selection exerts a large influence on results. Specification choices used to tweak causal models differ from their siblings used for the manipulation of observational studies. Causal models offer a different set of researcher degrees of freedom.

Why Results from Causal Models Can Be Tweaked

Causal models are inspired by the Rubin paradigm, which defines a causal effect as the difference in outcomes for treated units compared with a counterfactual outcome: the outcome in the absence of treatment for the units that have been treated (Rubin 1974; Holland 1986). Since the counterfactual does not exist—we can only observe the outcome given treatment for the treated group and the outcome given no treatment for the control group—different causal models have been developed that overcome this obstacle. Results from causal models can be tweaked in three main ways. First, tweakers can choose among different causal models or different subvariants of causal models. All causal models in one way or another work by selecting some cases and dropping others or dropping parts of the variation of each observation. Each one estimates what is known as a local average treatment effect, which will differ across causal models. Second, causal models are observational data analyses with a twist. They reduce but cannot eliminate the model dependence that comes with observational data analyses. Tweakers adopting causal model analyses of observational data can exploit that just as they can exploit the model dependence in traditional regression analyses of observational data. And third, tweakers can exploit the researcher degrees of freedom that come with uncertainty about the best model specification of each causal model. Causal models are model dependent (the name already provided a clue), though the model dependence is often hidden and difficult to uncover.

Tweakers exploit the researcher degrees of freedom that are specific to each causal model. Regression discontinuity design tweakers have different strategies and options than matching model tweakers, and both strategies vary from those of researchers who use instrumental variable models. Three causal models seek out credible counterfactuals by selecting some cases and discarding all others:

1. Matching (Rosenbaum and Rubin 1983) seeks to match treated and nontreated units so that they are as similar as possible on observed potentially confounding characteristics. Identification revolves around the unprovable claim that all important confounders are observable and have been matched on and that the balance in observed characteristics between treated and control group eliminates their potentially confounding impact on the treatment effect. Matching does not eliminate model dependence: researchers have to match on a model.
2. Regression discontinuity designs exploit settings in which some threshold separates those selected into treatment from those that are not selected into treatment.[1] Regression discontinuity designs require that a continuous observable running variable exists that determines selection.[2] This allows sampling cases close to the selection threshold—cases that either were selected into treatment (but almost not selected) or were not selected into treatment (but almost were selected). If the cases close to the selection threshold do not differ on potentially confounding factors, selection into treatment is quasi-random for cases within a certain bandwidth around the threshold.[3]

3. Difference-in-differences ("diff-in-diff") models are a variant of matching models that require time series data for a sample of cases. These research designs involve selecting cases that follow similar trajectories prior to the point at which some receive a treatment, while others do not. This is known as the parallel trend assumption for the cases in the selected sample. The identification assumption is that the treatment was quasi-randomly assigned by some force exogenous to the selected cases, an assumption that does not logically follow from the fact that treated and untreated cases were on similar trajectories because similar outcomes do not imply that the causes leading to that outcome were also similar.[4] Synthetic control is a variant of diff-in-diff: Rather than comparing actual existing discrete nontreated cases to treated cases, researchers synthetically create a weighted average of nontreated control units as a matched counterfactual to a treated unit. Typically, the synthetic control unit is created by matching on observable predictors of the outcome variable as well as the outcome itself before the treatment begins so that the artificially created control unit is as similar as possible on these observable pretreatment characteristics.

Two further causal models do not discard cases but instead discard some of the variation in variables across all cases in the sample.

1. The unit fixed effects model does not necessarily qualify as a causal model but is usually included in the discussion of causal models as it eliminates the potential for bias that results from the impact of confounders with time-invariant effects. This model discards all variation between cases and bases inferences on over-time variation within each case. Any claim to causal identification requires the strong assumption that the relevant confounding factors have constant, time-invariant effects on the outcome.
2. The instrumental variable model discards all variation in the treatment variable that is not correlated with the instrumental variable. Researchers assume that the treatment variable suffers from endogeneity either because it has been measured with systematic error, is affected by the outcome variable (reverse causality), or is correlated with confounding factors whose influence on the outcome cannot be appropriately controlled. Instruments eliminate bias only if the instrumental variable is uncorrelated with any model misspecification, if it does not affect the

outcome through any channel other than the treatment variable, and if the instrument affects the treatment variable sufficiently strongly.[5]

One might think that techniques claiming to identify the true causal effect of a treatment all estimate the same causal effect. But all causal models eliminate their very specific variance and they therefore estimate different causal effects. Causal models estimate what is known as the "local average treatment effect": the effect that holds for the individuals or units in the sample but, in the presence of causal heterogeneity, not necessarily for the population as a whole. The treatment effect "identified" is therefore the effect for a very specific and often quite narrow sample. By choosing one causal model over another, or one combination of models over another combination, tweakers can influence the quantity of interest in their favor. All causal models claim to estimate the *true causal effect* of a treatment on the outcome, but this truth varies between causal models and from one analysis to the other.

Of course, not all causal models are suitable to all research questions. Naturally, in the absence of selection into treatment, one cannot use regression discontinuity designs. If researchers do not analyze panel data, they can neither use diff-in-diff models nor drop the between variation with unit fixed effects. But matching and instruments are always an option. Thus, for virtually all research questions, researchers have a choice between identification strategies.

Identification strategies rely on distinct selection mechanisms. Thus, in the presence of causal heterogeneity, the local average treatment effects will be contingent on the specific identification approach employed. Throughout this book, we have repeatedly encountered fraudsters who manipulated empirical results by dropping selected cases. Wansink used sample selection strategies to find a publishable significant result. The same occurred in the InterMune case: The company sliced the sample of participants in the trials to find significant results.

The Wansink and InterMune cases *are* fraudulent. Yet few scientists consider identification strategies that carefully select some cases and drop others to be fraudulent. The difference, of course, is that Wansink and InterMune's Harkonen selected cases based on the empirical results of their remaining sample. And researchers who rely on causal models would never do that . . .

Each causal model reduces the model dependence of results that plagues traditional regression analysis. However, causal models do not reach independence from arbitrary assumptions. They only hide their model dependence better. Model dependence never disappears. At best it resurfaces in the specification choices required for matching cases; for selecting and specifying the running variable and bandwidth selection in regression discontinuity designs; for selecting the donor pool, observable variables, and weighting algorithm in synthetic control designs; or for selecting instrumental variables.

For *matching* models, there are numerous ways of matching treated and nontreated units, providing potentially myriad opportunities for tweakers. Recall that with options and choices comes tweakability. With causal heterogeneity, the estimated effect will depend on the set of cases selected into the matched sample. As Augurzky and Kluve (2007: 534) point out: "In implementing matching estimators, . . . many decisions—some of them ad hoc in nature—have to be made, and frequently it is not entirely clear if and how treatment effect estimates will be affected by these decisions." This provides fertile ground for tweakers. If coarsened exact matching does not work, use Mahalanobis distance matching, and if that does not work, use propensity score matching. If matching without replacement does not work, try matching with replacement instead. If exact matching does not work, try propensity score matching instead.

Propensity score matching, which is the most commonly used matching method despite King and Nielsen's (2019) recommendation that it should never be used, provides researchers with additional degrees of freedom because estimating propensity scores requires a model. However, this is true for all matching models: One cannot match without specifying a model based on which cases from the treatment and control groups are being matched.

If the set of confounding factors in a traditional regression analysis is uncertain, then the exact same uncertainty exists about which set of observables to match on in a matching model. Matching on one set of observables will produce a different matched sample and therefore different estimation results than matching on another set of observables. Tweakers can exploit this model dependence.

Ho et al. (2007: 199) argue for matching models by criticizing traditional regression models: "authors usually choose the presented estimates from numerous trial runs readers never see. . . . How do readers know that publications are not merely demonstrations that it is possible to find a specification that fits the author's favorite hypothesis?" Sure, but a claim that 2+2 is not 5 does not prove that it is 6. Accepted problems with one research design do not prove that any other design gives more reliable inferences, let alone finding the "truth." Yes, matching is meant as a method "for preprocessing data to improve causal inferences in observational data" (King and Nielsen 2019: 435). But nothing prevents a tweaker from changing matching algorithms after the estimation results have been observed, in order to "'improve" the results ex post in the researcher's interest.

Regression discontinuity models offer tweakers options in at least two dimensions. Researchers can choose the width of the selection corridor around the discontinuity, a fairly obvious tweak, and they can choose the functional form of the running variable that determines the selection into treatment, a less obvious tweak. The choice of the bandwidth follows a trade-off: The bandwidth has to be narrow to support the assumption that selection into treatment just above the discontinuity and out of treatment just below the discontinuity is quasi-random. At the same time, however, narrow bandwidths reduce the number of cases left in the sample. This not only leads to more inefficient estimates and a low external validity of the estimated or computed local average treatment effect. The choice of a corridor around the discontinuity is essentially at the researcher's discretion, which means that the selection of cases into the final sample is left to the researcher's discretion (Cattaneo and Vazquez-Bare 2017).

Similarly, the choice of the functional form of the running variable, which greatly influences the results in many applications, is at the researcher's discretion, and tweakers can choose one that shifts results in their interest. Are results more favorable for the tweaker when only the linear and its squared terms are included? Cite Andrew Gelman, a famous methodologist, and Nobel laureate Guido Imbens who advise against high-order polynomials (defined as at least third-order polynomials) because their inclusion leads to "noisy estimates, sensitivity to the degree of the polynomial, and poor coverage of confidence intervals" (Gelman and Imbens 2019: 447). Are the estimation results for the quantity of interest more favorable when higher-order polynomials are included? Cite one of the other methodologists who suggest choosing polynomials up to the order that gives the best model fit.

A review of the literature by Cattaneo and Titiunik (2022: 845) suggests that many of the issues surrounding the specification choices of regression discontinuity (RD) models have been resolved: "The choice of local polynomial bandwidth and local randomization neighborhood has become data driven and is now based on principled criteria, and many validation and falsification methods have been developed exploiting the particular features of RD designs. It is now understood that best practices for estimation and inference in RD designs should be based on tuning parameter choices that are objective, principled, and data driven, thereby removing researchers' ability to engage in arbitrary specification searching and leading to more credible and replicable empirical conclusions." Perhaps. But we still see publications in which both the bandwidths around the discontinuity and the functional form of the running variable are chosen ad hoc and without any demonstration of robustness. Likewise, "data-driven" does not have the same meaning as "optimal." Methodologists reduce researcher degrees of freedom in regression discontinuity designs by recommending data-driven specifications, but this strategy comes at the cost of potentially suboptimal model specifications. Reporting just the one "data-driven" specification also effectively prevents robustness tests that explore whether results are robust. If estimated effect sizes are unstable and not robust to small changes in model specifications, researchers should not claim that they have identified *the* true causal effect. Not exploring the robustness of one's results generates overconfidence in one's results.

Difference-in-differences models afford researchers discretion over which cases to include in the sample. Reviewers will expect to see evidence that the parallel trend assumption holds, but researchers are free to choose cases within that requirement. Researcher degrees of freedom proliferate with the synthetic control variant of difference-in-differences. Athey and Imbens (2017: 9) praise synthetic control as "arguably the most important innovation in the policy evaluation literature in the last 15 years. This method builds on difference-in-differences estimation, but uses systematically more attractive comparisons." That was not what Athey and Imbens had in mind when they wrote this sentence, but tweakers will agree that synthetic control is more attractive than simple difference-in-differences. Researchers control which cases go into the pool from which the synthetic control group is constructed. Researchers also influence the set of observable variables and the matching algorithm on observables that provides the weights with which some of the donors are drawn from the pool. If one synthetic control counterfactual unit does not produce the desired result, tweakers can go on the hunt for another one that does.

With *instrumental variable* (IV) models, tweakability arises because different instruments generate different results, and tweakers can choose the instrument that produces the desired result.[6] By regressing the treatment variable on the instrument and all other explanatory variables, they use the predicted variation in the treatment variable from this "first stage" regression in the "second stage" estimation model. One instrument will produce one treatment effect estimate and another instrument another (Angrist 2004). This is excellent news for tweakers.

Instrument validity has to be assumed and argued for; it cannot be proven. What is a major drawback for causal inference and scientific progress is good news for tweakers. Tweakers ought to love instrumental variable models because they can simply try out different instruments, choose the one that produces the most desired result for the quantity of interest, and then tell a story about why this instrument is plausibly valid. Since one instrument will produce one local average treatment effect and another instrument will produce another such effect, the chances are that a skilled tweaker can produce almost any desired effect by choosing their favorite instrument. In the extreme, researchers could, if they wanted, produce a desired result that is in the opposite direction of the true causal effect. As Heckman and Urzua (2010: 34) explain: "Most IV studies do not clearly formulate the economic question being answered by the IV analysis. . . . In the binary outcome case, even if Z is a valid instrument, if Z is a vector, and analysts use only one component of the vector as an instrument, and do not condition on the other components of Z, the weights on the MTE [marginal treatment effect] can be negative over certain ranges. The practice of not conditioning on the other instruments is common in the literature. In this case, IV can estimate the wrong sign for the true causal effect."

Tweakers do not necessarily have to find all possible instruments and then choose the optimal one for their purposes. Instead, they try out promising potential instruments until they have found one that produces the desired results. The larger the set of potential instruments to choose from, the greater the influence researchers have on the estimation result. For tweakers, the greater the heterogeneity in the effects resulting from each of the available instruments, the better. If multiple instruments produce quite different results, then either this is because they are not valid instruments, which cannot be tested, or, assuming they are all valid, this demonstrates causal heterogeneity as each one represents a different local average treatment effect. Tweakers pick one and then claim to have "identified" the true causal effect of the treatment variable on the outcome.

How Much Tweaking Is There in Causal Models?

Cherry-picking is a problem for all research designs, not just for causal models. Researchers relying on these models have the same interest as any other scientist: They do not want to waste a potentially promising paper because of a lack of statistical significance or some other inconvenience with the results.

Abel Brodeur, Nikolai Cook, and Anthony Heyes's study titled "P-Hacking and Publication Bias in Causal Analysis in Economics," published in the *American Economic Review* (Brodeur et al. 2020), focuses on multiple causal models. They analyze results of the testing of 21,000 hypotheses published in twenty-five leading economics journals and use three different approaches to estimate the proportion of p-hacked causal analyses.[7] P-hacking analyses provide evidence for tweaking at the aggregate level. However, they cannot identify which individual researchers have tweaked the p-values of their studies. Note that the methods used by Brodeur and co-authors do not capture tweaks that have a large effect on p-values, moving them further away from the conventional threshold than just below. Nevertheless, they find that tweaking does not appear to be uncommon in causal models in economics. The results are consistent across the three different approaches and suggest that around 15 percent of results in instrumental variables and difference-in-differences models and around 1.5 percent of results in regression discontinuity models have been p-hacked. Interestingly, Brodeur and co-authors found no evidence that papers published in one of the "top five"' economics journals were any less guilty of p-hacking than those in other journals nor that the situation improved over time.

These results are broadly in agreement with our expectation that tweaks in instrumental variables and difference-in-differences models are easy to implement. Results are also consistent with Brodeur et al.'s (2020: 3636) own assessment that the different degrees of freedom offered by different models can explain the uneven preponderance of p-hacking across causal models. However, another study that addresses the reliability of findings from regression discontinuity models, albeit in political science rather than economics publications, suggests that Brodeur et al. (2020) underestimate the extent of tweaking with these models. Evaluating all regression discontinuity designs published in three top general political science journals between 2009 and 2018, Stommes et al. (2021: 12) come to the following sobering conclusion: "The body of published political science research using the RD design exhibits pathological features consistent with selective reporting or publishing. More than half of the studies in our sample do not properly estimate the accuracy of their estimates, leading to the associated hypothesis tests and confidence intervals being misleading. Most of the studies are underpowered, and are able to detect only large effects. Taken together, this paints a somber picture of the state of applied studies using the regression discontinuity design. Our results suggest that many published findings using the design are exaggerated if not altogether spurious." So much for claims of causal identification.

Stommes et al. (2021) believe that their findings are not driven by researcher degrees of freedom. They base this verdict on finding no real differences when comparing the prevalence of pathological features in regression discontinuity models with automated bandwidth selection procedures to those with nonautomated procedures, where researchers have more discretion. We do not think that allowing an algorithm to determine the bandwidths is a good idea, nor do we believe that this strategy provides a safeguard against tweaking through bandwidth selection. There are certainly other ways to tweak regression discontinuity designs. Most of the replicated studies were not well powered and therefore stood little chance of detecting small effects with their designs. They rely, in one way or another, on throwing away information.

Throwing away variation comes with real cost. Empirical researchers often assume that discarding information merely reduces efficiency. They believe that as the amount of information decreases, standard errors become larger and that is it. Indeed, standard errors usually become larger when the sample size declines. However, the assumption that nothing else happens is false. Inefficient estimation can lead to point estimates that are far from the truth even if the standard errors remain moderate or small. As Gelman and Carlin (2014: 649) caution: "There is a common misperception that if you happen to obtain statistical significance with low power, then you have achieved a particularly impressive feat, obtaining scientific success under difficult conditions. However, that is incorrect if the goal is scientific understanding rather than (say) publication in a top journal. In fact, significant results in a noisy setting are highly likely to be in the wrong direction and invariably overestimate the absolute values of any actual effect sizes, often by a substantial factor."

Tweakers can capitalize on erroneous results in causal models that have a low signal-to-noise ratio, and this should concern anyone worried about tweaking. It is not uncommon for regression discontinuity models or estimations based on exact matching to reduce the number of cases by 90 percent. The instrumental variable model is a notorious example showing that the problem can be serious. Even in the unlikely event that researchers rely on a perfectly valid instrument, instrumental variable models are just consistent; they are not unbiased in finite samples. In small- to medium-sized samples the estimate often deviates substantially from the true effect. Angrist and Krueger (2001: 71) rightly complain that textbooks sometimes gloss over the distinction between unbiasedness and consistency.

Similar problems arise with weak instruments. Estimates with weak instruments are especially likely to be biased. Such estimates are therefore not necessarily closer to the truth than estimates from traditional regression analyses. Young (2022: 2) reanalyzes more than a thousand instrumental variable regressions in thirty articles published in the journals of the American Economic Association. He comes to the sobering conclusion that instrumental variable regression "may be prized for its asymptotic consistency, but in finite samples it often allows for very little inference."

The reduction in the effectively used information and hence in the signal-to-noise ratio can be influenced by the researcher by choosing one causal model over another or by adopting one specification choice over another within a particular causal model. Suppose that, purely by chance, the correlation between the error variance that deviates from the assumed standard normal distribution of the residuals and the treatment variable is large. In this not entirely unlikely situation, the more information the researcher omits, the greater the impact of noise on the estimate. If the correlation between noise and treatment pushes the coefficient in the desired direction, researchers occasionally improve the publishability of their work by increasing the influence of errors on the estimates.

Does an Extra $100 Improve Learning Success?

With a financial injection of $138 million spread across some 1,000 schools and benefiting 1.4 million students, the question naturally arises: Can such funding have a discernible impact on educational outcomes? According to an analysis by Kristian Holden (2016), the answer is yes. Holden claims that allocating funds for textbooks has a significant and positive impact on student achievement. Holden's results tell a seemingly compelling story, in which textbooks prove to be a remarkably cost-effective strategy for raising average student test scores compared with other interventions such as class size adjustment.

Holden (2016) relies on a fairly standard regression discontinuity design. The analysis is transparently documented with replication data and do-files publicly available.[8] The regression discontinuity design exploits a discontinuity in school funding that was created by a settlement after a protracted legal battle against the state of California. The initiative, known as the *Instructional Materials—Williams Case* (IMWC), took its name from the plaintiff who spearheaded the "Williams settlement" with California. After four years of litigation, the state agreed to provide all schools with some extra funding that increases over the years from $25 per student to $54.22 per student. In addition, the lowest-performing 20 percent of the schools received an additional one-off payment of $96.90 per student for textbooks. Holden exploits this additional one-off payment for his regression discontinuity design.

Holden (2016: 123) claims to have "identified the effect of textbook funding on student achievement by focusing on a quasi-experiment created by the Williams settlement." Interpretations like this are fairly common in causal analyses, but they are bold overgeneralizations. At best, Holden has shown that students in California elementary schools that received book subsidies performed slightly better than students from schools that did not receive book subsidiaries. Holden has not shown whether these results carry over to schools further away from the eligibility threshold, whether these effects exist outside California in other US states or in any other year. It doesn't take a deep dive into the economics of education to see that the impact of subsidized textbooks on learning outcomes is inherently and necessarily heterogeneous, varying between students, schools, locations, and over time. Researchers employing causal models commonly overgeneralize their empirical findings, and Holden is no exception. Importantly, a methodological tool that allows researchers to generalize inferences based on causal analyses to the population does not exist.

We use Holden's analysis as an example that the results of regression discontinuity designs can be tweaked and tweaked with ease. We do not know whether Holden had actually tweaked his results. Our reanalysis suggests that the causal effect of schoolbook subsidies is smaller and arguably less general than Holden suggests. But that does not imply, and certainly does not prove, that Holden had actually tweaked his result. Our modest objective is to show how replacing many of the model specification assumptions in regression discontinuity design by other, at least equally plausible assumptions has, or at least can have, strong effects on predicted causal effects. In Holden's case, the vast majority of plausible alternative specifications give smaller and often less significant results. In effect, such results would have been difficult to publish. Again, nothing proves tweaking, but we would not bet a fortune that Holden's results replicate in a perfect experimental design that randomizes eligible schools.

Tweak 1: The Dependent Variable

Holden (2016) analyses three standardized average student test scores: a math score, a reading score, and the unweighted average of these two scores. The running variable is the 2003 Academic Performance Index (API). Table 7.1 displays results based on linear regression with robust standard errors, representing an exact replication of Holden for the treatment effect in the year 2005 when the effect was at its strongest.

Table 7.1
Effect Size Estimation with Different "Dynamic" Specifications

	Exact replication	Absolute change	Relative change	Lagged dependent variable
Eligibility threshold	API = 643	API = 643	API = 643	API = 643
lagged dep. variable				.866 (.051)
treatment	.203 (.056)	.108 (.056)	−3.606 (2.839)	.120 (.054)
API (treated)	.018 (.004)	.006 (.004)	−.157 (.140)	.008 (.004)
API (nontreated)	.007 (.006)	.002 (.042)	.088 (.150)	.003 (.005)
N	536	536	536	536
R^2	.120	.016	.004	.464
Lower bound	> 623	> 623	> 623	> 623
Treated	< 644	< 644	< 644	< 644
Upper bound	< 663	< 663	< 663	< 663

Defining the dependent variable as the average standardized test scores of students in a school is a specification choice: one that does not fit the theory perfectly. Holden is interested in the effect of the textbook subsidy treatment on student achievement. The best research design to test this theoretical prediction operationalizes the dependent variable as the absolute change in achievement scores from the pretreatment to the posttreatment years, controlling for the lagged change in achievement scores. One could also analyze the relative change, the growth rate of the performance scores. Both specifications are close to the theoretical logic underlying the treatment effect of textbook subsidies, and both specifications more closely fit the data-generating process than Holden's preferred specification does, since school performance data show a strong temporal correlation. Not surprisingly, math and reading skills in one year strongly depend on math and reading skills of the previous years. The autocorrelation coefficient is about 0.78 and thus way too large to ignore.

Table 7.1 presents additional results from models with these two alternative dependent variables, one the absolute change in students' average standardized test scores and the other the relative change or growth rate in test scores. Controlling for the autocorrelation in the data exerts a strong influence on results. The level of statistical significance also declines to levels just below conventional thresholds. Results get worse in the model using the growth rate in performance as dependent variable: Any evidence of a positive treatment effect evaporates. Results from the model with the lagged dependent variable are close to those with absolute change as dependent variable.

A closer inspection of the data also reveals that schools' performance levels converge. And they converge independently of schoolbook subsidies. Table 7.2 displays the 2005–2011 averages of the levels and absolute annual change of the mathematics score and the reading score for each decile of the 2003 API score.

The average change in test score results over the period 2005–2011 is a function of the 2003 API score. The higher the decile on the 2003 API score, the smaller the absolute change over time. This offers an alternative interpretation for the observed variance in test scores: Schools in the third decile that did not receive schoolbook subsidies experience a smaller annual improvement in test scores than schools in decile 2 that did receive schoolbook subsidies, not necessarily because of these subsidies, the treatment, but because they initially had higher test scores and because of the factors that cause the convergence in test scores over time. This alternative

Table 7.2
Level and Absolute Change in 2005–2011 Test Scores per API Decile

	Math score		Reading score	
Decile	Level	Change	Level	Change
1	–0.251	0.302	–0.578	0.281
2	–0.069	0.279	–0.256	0.264
3	–0.005	0.245	–0.132	0.245
4	0.268	0.270	0.117	0.230
5	0.318	0.237	0.265	0.230
6	0.428	0.220	0.405	0.215
7	0.555	0.223	0.561	0.197
8	0.664	0.210	0.761	0.197
9	0.819	0.214	0.923	0.195
10	1.012	0.214	1.165	0.195

explanation does not imply that schoolbook subsidies have no effect, but it suggests that researchers ought to control for convergence of schools in the analysis.

The standard model specification that accounts for convergence and temporal dynamics includes the test score from the previous year. These specifications are appropriately known as lagged dependent variable models. With a coefficient of 0.866 for the lagged dependent variable, the performance of schools in 2005 is closely related to their performance in 2004, a feature of the data-generating process Holden ignores in his analysis. And this omission matters: Compared with Holden's preferred specification, the size of the predicted treatment effect declines by roughly 40 percent.

Tweak 2: The Eligibility Threshold

The state of California implementing the Williams settlement set the eligibility threshold at the bottom two deciles of the published 2003 API database, which correspond to the exact thresholds used by Holden. This is not exactly the upper bound of the second lowest decile in the actual sample analyzed by Holden. The eligibility threshold for textbook subsidies was set by the California state government and school boards. Nevertheless, it is useful to test the robustness of the results to apparently small changes in the eligibility threshold.

Table 7.3 illustrates the influence that a placebo increase in the eligibility threshold has on the predicted treatment effects. These minor adjustments

Table 7.3
Effect Size Estimation with Various Eligibility Thresholds

	Exact replication			
Eligibility threshold	API = 643	API = 644	API = 645	API = 646
treatment	.203 (.056)	.122 (.073)	.127 (.072)	.125 (.073)
API (treated)	.018 (.004)	.015 (.004)	.017 (.004)	.021 (.005)
API (nontreated)	.007 (.006)	.005 (.006)	.001 (.053)	−.007 (.055)
N	536	542	544	543
R^2	.120	.109	.109	.109
Lower bound	> 623	> 624	> 625	> 626
Treated	< 644	< 645	< 646	< 647
Upper bound	< 663	< 664	< 665	< 666

lead to a notable reduction in the predicted size of the treatment effect. The standard error increases, eliminating the statistical significance of the predicted treatment effect.

To be clear: The estimates at eligibility thresholds higher than 643 are based on incorrect model specifications since schools with an API higher than 643 did not receive the additional funding. Our robustness test follows the logic of a placebo test. However, the placebo is not very different from the true treatment, and thus we would expect that the predicted effect is relatively stable. Table 7.3 demonstrates just how fragile results are with respect to the inclusion in the sample or exclusion from the sample of a very small number of schools as well as their correct classification to below or above the threshold. Results suggest that Holden's findings are driven by a small number of observations. Another way to show this is to ask how many schools we would have to selectively and purposely remove from Holden's sample to render the treatment effect either substantively small or statistically insignificant. We find that we need to drop only twelve schools from Holden's sample, a sample that covers 2,350 elementary schools in total and that Holden reduces to 536.

Tweak 3: Sample Preselection

The perhaps least visible tweak of regression discontinuity analyses preselects observations by dropping some of the observations that fall outside the bandwidths: schools with either very low or very high performance scores. Given modern computer technology, this sample preselection has no obvious advantage. Eliminating schools that are not close to the eligibility threshold appears harmless: These schools will not fall within the bandwidth that determines the sample. What difference could their exclusion possibly make?

The simple truth is that preselection of cases makes a huge difference for the results reported by Holden. The preselection of cases influences sample properties such as the range, the mean, and the variance of the cases included in the sample. Preselection thus affects estimation results if sample properties are used to standardize variables. And that is what Holden does: He standardized schools' performance indicators by the sample variance of these performance indicators.

Holden preselects only schools that are at most 100 API scores below or above the eligibility threshold of 643. Holden never justifies his sample preselection decisions, and we do not see any good theoretical or empirical justification to manipulate the sample variance used to standardize

Table 7.4
Effect Size Estimation with Different Case Selection Corridors

			Exact replication		
Preselection corridor	None	493..793	543..743	593..693	624..663
treatment	.131 (.047)	.185 (.066)	.203 (.073)	.222 (.079)	.227 (.081)
API (treated)	.016 (.003)	.023 (.004)	.025 (.005)	.027 (.005)	.028 (.005)
API (nontreated)	.011 (.003)	.026 (.004)	.018 (.004)	.019 (.005)	.020 (.005)
N	536	536	536	536	536
R^2	.119	.119	.119	.120	.120
Lower bound	> 623	> 623	> 623	> 623	> 623
Treated	< 644	< 644	< 644	< 644	< 644
Upper bound	< 663	< 663	< 663	< 663	< 663

performance indicators. If we base the test score standardization on the full sample and re-estimate his baseline model, we obtain a treatment effect that is roughly one third smaller than the effect reported by Holden. Other model specifications produce larger substantive effect sizes than those reported by Holden. Preselecting the sample to observations with an API score between 593 and 693 gives effect size estimates that are approximately 10 percent larger than the one reported by Holden (2016). But this comes at the price of significantly larger standard errors, a price that tweakers may not be willing to pay.

Tweak 4: Choice of Bandwidths

In regression discontinuity designs, the choice of a bandwidth is a crucial decision. Choosing the bandwidth is an obvious tweaking option—so obvious that many researchers report robustness tests with different bands to preempt concerns by reviewers or readers.

Holden bases his choice of a symmetric bandwidth on what methodologists have called the "optimal selection procedure" (Calonico et al. 2014). The chosen bandwidth minimizes the mean squared error of the model. But this is an arbitrary criterion. Just because the specification criteria follow a rule does not mean that there are no alternatives and that the one chosen is the "right" one, which it is not. We change the widths of Holden's chosen bands in relatively small steps. However, these small steps have a

Table 7.5a
Variation of Upper and Lower Thresholds

			Exact replication	
Treatment	.062	.118	.203	.180
	(.052)	(.060)	(.056)	(.101)
API score	0.012	.015	.018	.018
	(.001)	(.003)	(.004)	(.015)
Interaction	.002	.002	.007	-.001
	(.002)	(.003)	(.006)	(.019)
N	1065	796	536	277
R^2	.303	.214	.006	.012
Lower bound	> 603	> 613	> 623	> 633
Treated	< 644	< 644	< 644	< 644
Upper bound	< 683	< 673	< 663	< 653

Table 7.5b
Variation of Lower Threshold

			Exact replication	
Treatment	.120 (.064)	.144 (.067)	.203 (.056)	.178 (.082)
API score	0.018 (.004)	.018 (.004)	.018 (.004)	.018 (.004)
Interaction	–.004 (.005)	–.001 (.005)	.007 (.006)	.000 (.013)
N	758	668	536	418
R^2	.204	.214	.006	.044
Lower bound	> 603	> 613	> 623	> 633
Treated	< 644	< 644	< 644	< 644
Upper bound	< 663	< 663	< 663	< 663

Table 7.5c
Variation of Upper Threshold

			Exact replication	
Treatment	.146 (.062)	.177 (.066)	.203 (.056)	.205 (.094)
API score	0.012 (.001)	.015 (.003)	.018 (.004)	.019 (.015)
interaction	.013 (.005)	.010 (.005)	.007 (.006)	.006 (.015)
N	843	664	536	394
R^2	.199	.159	.006	.084
Lower bound	> 623	> 623	> 623	> 623
Treated	< 644	< 644	< 644	< 644
Upper bound	< 683	< 673	< 663	< 653

large effect on the predicted treatment effect at the discontinuity. It does not take much to make the treatment effect much smaller or, worse for publication, to eliminate the statistical significance of the treatment effect.

With only one exception the point estimates are always smaller than the one reported by Holden and are often no longer statistically significant. The "optimized" bandwidth choice of Holden (2016), coincidentally or not, gives the "best," the most publishable combination of effect size and significance. Regression discontinuity design can be shown to be vulnerable to changes in bandwidth specification assumptions.

Tweak 5: The Functional Form of the Running Variable

The results of regression discontinuity designs are (almost) inevitably influenced by assumptions made regarding the functional form of the running variable. Holden assumes that the functional form that relates schools' API scores to their performance scores is linear. This assumption runs counter to economic theory, which sees diminishing returns everywhere. And indeed, a simple scatterplot relating API scores to average performance indices shows a moderately convex function—a function perfectly consistent with the theory of diminishing returns but not so with the linearity by default assumption Holden uses.

Thus, our next exploratory test of the tweakability of this particular RD design replaces the linearity assumption of the running variable with two perhaps equally arbitrary nonlinear functional form assumptions (table 7.6). These results add more evidence, if needed, that the results of a regression discontinuity design can be sensitive to basic specification choices. RD designs are, like all other causal models, not independent of modeling choices. Perhaps to the contrary: Causal models offer a myriad of seemingly innocuous specification assumptions, and plausible alternatives to

Table 7.6
Effect Size Estimation with Different Functional Form Assumptions

	Exact replication	3rd polynomial	3rd and 5th polynomial
Treatment	.203 (.073)	.165 (.094)	.171 (.111)
APIs of treated and untreated (included controls)	API	API, API^3	API, API^3, API^5
N	536	536	536
R^2	.121	.121	.121
Lower bound	> 623	> 623	> 623
Treated	< 644	< 644	< 644
Upper bound	< 663	< 663	< 663

the assumptions chosen tend to come with relevant changes to estimated treatment effects.

We can further display the intricate relationship between API scores and average school performance by using a non-parametric modeling approach. This approach eliminates the constraints of parametric models. Results are shown in figure 7.1. The dashed lines represent the bandwidth chosen by Holden.

The predicted effect size from Holden's regression discontinuity design results from the assumed boundary and, more importantly, the choice of lower and upper bounds. A cursory glance reveals that the slope of the function is greatest between an API of 623 and one of 643. It is heroic to generalize the larger slope for schools with an API rank between 623 and 643 to all other schools. Some students and their schools benefited from the textbook subsidy program. The evidence that all treated schools benefited is weak at best.

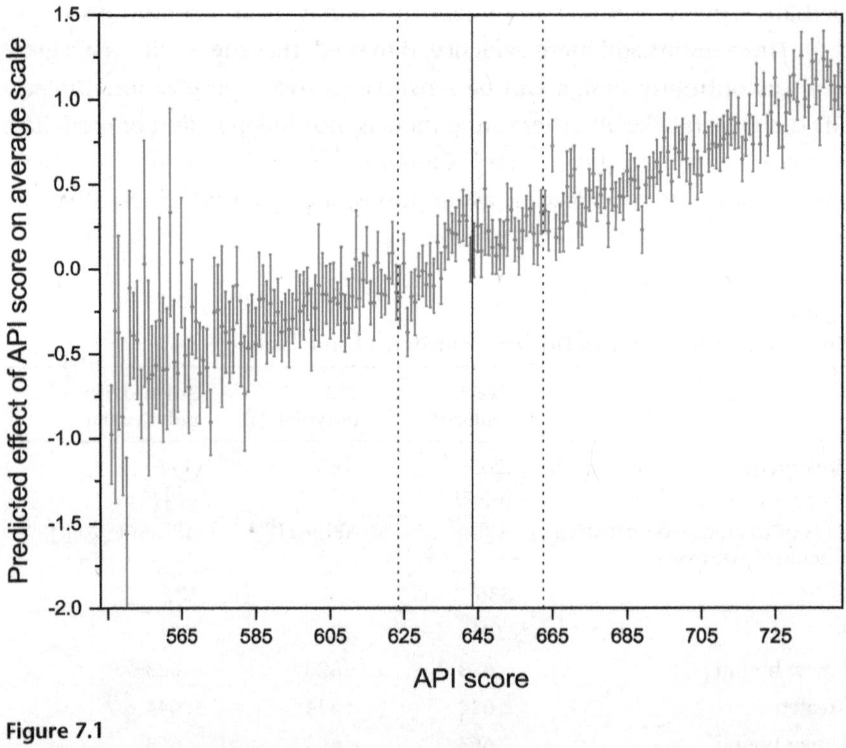

Figure 7.1
Predicted effect sizes of API levels on school performance: semiparametric model.

Tweak 6: Subsample Analysis

The Williams settlement provided a one-off additional payment not only to low-performing elementary schools but to all schools, and thus also to middle and high schools. The eligibility thresholds are different for these other two types of schools, set at their respective upper boundary of their bottom two deciles, namely, API scores of 600 and 584, respectively. One would therefore expect estimation results to be based on the sample containing all schools.

Holden started out with a full sample of all schools. Yet he never reports results based on a full sample. We can only speculate why, but one potential reason is that results are not as favorable as for the subsample of elementary schools only. For example, the 2005 treatment effect for all schools is small and not significant at conventional levels.

Holden instead reports results separately for middle and higher school subsamples for which he finds insignificant or, in some cases, "wrongly" signed effects. Holden's explanation for the results for middle and higher schools is unpersuasive. He suggests that they are due to there being far fewer middle and high schools: "In California, there are roughly five times as many elementary schools than high schools or middle schools. . . . As a result, the standard errors of the estimated effects are very large and we cannot rule out economically significant effects on middle schools or high schools" (Holden 2016: 122). True, sample sizes influence the level of statistical significance. However, the problem does not result only from larger standard errors. All estimates for middle schools are wrongly signed. The treatment effect in the year 2005 is positive for higher schools and for elementary schools, but it is much smaller, only roughly one-third the size it is estimated to have for elementary schools, and with standard errors almost four times the size of the coefficient, the estimated effects are miles away from significance. Hence, Holden finds a clear causal heterogeneity between school types, a heterogeneity that he does not really explain. This shows how, if the aggregate estimation results are not favorable, tweakers can exploit subsample analysis to find estimates in subsamples that are favorable. This is a popular tweaking strategy in other research designs but of course also an option in causal models.

Discussion

Regression discontinuity designs offer ample opportunities to tweak empirical results. All the researcher has to do is search, which, given the power of modern computers, is a matter of hours, and, of course, one can let the computer do the hard work of estimating thousands of models with slightly different specifications to find the "right" results: those results that promise a publication.

Using Holden's analysis of textbook subsidies as an example, we demonstrated how much the empirical results of his regression discontinuity design can be tweaked. Regression discontinuity designs are easily tweaked to an extent that makes all the difference between unpublishable and publishable empirical results. We see no reason to think that what applies to regression discontinuity designs does not also apply to other causal models.

To be sure, we do not know whether Holden has actually tweaked. But we do know that he could have. Others have commented that the core reported substantive effect size is very large, perhaps surprisingly so. Kirabo Jackson (2020: 10), to take one example, is a little puzzled by the size of the effect Holden finds but rushes to offer an explanation: "This estimated effect is larger than most in the literature. However, this could potentially be rationalized by a model in which spending on textbooks was inefficiently low." Perhaps. Or perhaps Holden has inflated the predicted treatment effect by employing one or more of the tweaking options we discussed or other ones we have not. That Holden's choice of model specification, deliberately or not, appears to maximize a combination of effect size and significance level raises the suspicion that Holden may have chosen the specification based on the results. But suspicion is not proof.

Conclusion

According to their proponents, causal models have improved the validity of inferences. Maybe that is occasionally true. But the rise of causal models has opened the floodgates for tweakers who love few things more than specification choices that give them options for getting the estimates they want and need. Causal models add more options to the tweaker's toolkit.

Researcher degrees of freedom loom large in causal models. The nature of causal models is the nonrandom selection of cases or variances, and since, at least in the social sciences, effect sizes are conditional and depend on analyzed cases, the permission to select cases is identical to the permission to select the effect size. Specification choices exist in all causal models. Causal models invite researchers to actively select cases or the variance they use to estimate what they then call the "true causal effect," a term that should be banned from serious scientific publications.

Only reanalyses can tell whether the results of causal models are replicable and robust to alternative plausible model specification choices. Not much follows from the identification of "a" causal effect. As with any other research design, little confidence can be placed in the results of any single causal model even in the absence of tweaking. Only if a sufficiently large number of independent empirical analyses arrive at the same or similar conclusion, the same or similar inference, can we begin to believe that results are robust and, perhaps, valid. Alternatively, if they reach different conclusions and different inferences, then either the restrictive assumptions on which their claims to inferential validity are based do not hold or they simply uncover and reveal causal heterogeneity. What holds in the absence of tweaking holds even more if at least some researchers tweak at least some of the time: The results of any single causal model are not a solid ground for making strong inferences. Such is the nature of local average treatment effects, such is the potential for bias in these studies, and such is the logic of tweaking.

Much more appreciation is needed for the fact that some model dependence from traditional regression analysis remains while other specification choices unique to causal models introduce new ways of tweakability. But this appreciation is in short supply and hard to come by when the advocates and users of causal models adopt their triumphalist and condescending tone, claiming to be quasi-experimentalists operating far away from the darkness of traditional regression analyses that produce mere "pseudo-general pseudo-facts" (Samii 2016: 941). Nobel Prize winner James J. Heckman is forensic in his exposition and scathing critique of how some causal models can be used to produce anything but "credible" inferences. As early as 2010, Heckman and Urzua (2010: 35) warned: "Since, in the general case, different instruments identify different parameters, IV is not robust to the choice of instrument. Since the sign of an IV can be different from the true causal effect, IV models may at times produce a misleading guide to policy or inference, so it is not robust." Young (2022: 20), based on insights from his reanalysis and evaluation of instrumental variable models published in top economics journals, similarly cautions his fellow economists against methodological triumphalism: "Learning about the world may simply be harder than suggested by simple dichotomies between good and bad research designs." As more studies like Brodeur et al. (2022) and Stommes et al. (2021) are published demonstrating p-hacking, publication bias, or lack of reliability of causal models, the tide will turn further and more awareness of the tweakability of causal models will sink in.

III Countering Fraud

> For science to work, it has to depend on more than just the honesty of its individual practitioners.
>
> —Lee McIntyre

The West Yorkshire town of Bradford is the UK's third most dangerous city, with nearly double the average crime rate in England, Wales, and Northern Ireland. To improve the situation, the West Yorkshire Police Department maintains an informative web page on principles of crime prevention.[1] Given the high crime rate in and around Bradford, one might conclude that this web page and the county's underlying crime prevention policies on which it is based do not help much, but such a conclusion is obviously unwarranted: We do not know the counterfactual crime rate in West Yorkshire in the absence of the crime prevention recommendations. One thing we do know is that the county's crime prevention strategy rests on three pillars that can without much stretch be applied to scientific fraud. These three pillars are deterrence, prevention, and detection. Not surprisingly, suggestions to counter scientific fraud also rest on these three pillars, though the classification of the various strategies may appear slightly arbitrary. Whistleblowing, for example, can be perceived as a deterrent strategy, but whistleblowers are also one of the main reasons why fraudsters are detected.

Ideas for deterring scientific fraud include proposals such as severe sanctions and criminalization of fraudsters. These strategies have been tried with little apparent effect, but we do not know the counterfactual level of fraud in the absence of sanctions. Nevertheless, it seems plausible to assume that just as some criminals are not deterred by the police, not all fraudsters can be deterred by harsh punishment of detected fraudsters. At the other end of the spectrum of potential solutions, we see strategies that level the playing field (legalize it!) to strategies that try to disincentivize fraud, removing the close link between publications and academic recognition, promotions, and fame.

Virtually all promising strategies for reducing scientific fraud depend on substantial improvements in detection rates. If detection rates remain low, the incentive to publish fraudulent research will remain high. And what is the value of criminalization if detection is overwhelmingly random and unsystematic? Although the detection strategies available to researchers have improved and include forensic statistical scrutiny of results, replication and robustness testing, and artificial intelligence-based red flags for manipulation and duplication, detection rates remain low. In addition, detection strategies face the insurmountable problem that tweaks are logically indistinguishable from specification searches. Currently, detection techniques fail to reliably identify fraudulent publications. The false positive rate is not zero, and the false negative rate remains unacceptably high.

8 Deterrence Policies

> When I started to fake my studies, it was very, very easy. I was always alone. No one checked my work; everyone trusted me. I did everything myself. Everything. I thought up the idea, I did the experiment, and I evaluated the results. I was judge, jury, and executioner.
>
> —Diederik Stapel

Defending Trust

The ivory tower of science is built on trust. Scientists stand on the shoulders of giants and, occasionally, of dwarfs, too. But standing on the shoulders, building on the work of others, is what they do. Scientists must have faith that the shoulders they stand on will hold them up and carry them forward. If researchers cheat by fabricating or manipulating data or tweaking analyses, those standing on their shoulders see their research compromised and their careers potentially damaged, though they have not done anything wrong other than rely on the wrong published work.

Trust is not only needed between scientists who cite each other's work. Scientists want and need the general public to trust them as well. According to Sudip Parikh, chief executive officer of the American Association for the Advancement of Science and editor-in-chief of the *Science* family of journals, there exists no obvious alternative to trust because laypeople have no chance of critically interrogating twenty-first-century science, with its technical jargon.[1] Since the public cannot fully comprehend modern scientific contributions, public trust in scientists in general is a prerequisite for trust in scientific results and for public support of science as an enterprise. At the very least, scientists need taxpayers' money to fund their research.

It does not suffice to build trust and then forget about it. Earning trust is hard work, maintaining trust is challenging, and losing it is easy. Scientific misconduct in all its facets undermines trust and leads to its erosion. This holds true for major wrongdoing such as data fabrication, as well as for minor wrongdoing, such as tweaking. Science has recently been in the news with reports on world-famous fraudsters, the replication crisis in psychology and other disciplines, and Ioannidis's (2005) headline-grabbing claim that most published research results are false. No wonder that the public can no longer distinguish charlatans from honest researchers.

Once trust has eroded, the fabric of science can only be held together by explicit rules, effective monitoring of research practices, and a reasonably effective enforcement of sanctions in cases of misconduct. In principle, the self-governance of science by and for scientists can achieve this. Public shaming of fraudsters, the official retraction of their papers, and their expulsion from the scientific community will deter some. However, too many fraudsters are clearly not deterred by this risk, and, almost by definition, mere tweakers cannot be deterred since their tweaking remains hidden in virtually all cases. Where self-policing by scientists fails or remains ineffective, the institutions that employ them are unlikely to step in and come to the rescue. University and research institute leaders have preferences that usually align with their researchers. They too seek more publications, more research funding, more fame. The Retraction Watch website provides ample evidence that universities respond too little, too late, when one of their researchers is suspected of fraud.

Other institutions within the scientific community, most notably academic journals and funding agencies, are trying to step in. Journals will investigate if allegations are brought to them that an article they have published is based on fraud and will retract articles whose authors are found guilty. Funding agencies do not wish to get involved in allegations of malpractice and instead delegate responsibility to the research institutions receiving the funding. For example, the Deutsche Forschungsgemeinschaft, the German equivalent of the UK Research and Innovation department or the American National Science Foundation, states the following in its "Guidelines for Safeguarding Good Research Practice—Code of Conduct": "The heads of [Higher Education Institutions] and non-HEI research institutions . . . are responsible for ensuring adherence to and the promotion of good practice. The heads of research institutions guarantee the necessary conditions to enable researchers to comply with legal and ethical standards."[2] Easier said than done: each player in the scientific ecosystem has an incentive to shift the burden of maintaining trust in science to others. Fraud deterrence constitutes a severe collective action problem. This may explain why many institutions currently try to motivate others to solve the problem.

Self-Correction

Science is a communicative and cumulative process based on conjectures and theories that are empirically tested, often by the very same scholars who formulated the conjectures and theories, as well as by others. Empirical research often seeks to provide evidence to support (or refute) a theoretical idea, but evidence can also be used inductively to develop new ideas and to challenge established theories. For all its flaws, science often quickly identifies and rejects theories that are not supported by empirical evidence. Science self-corrects some false inferences over time, including those that result from tweaking or outright fraud. Fraudsters who are caught suffer public shame, resulting in a loss of reputation and sometimes their jobs.

Many commentators on scientific fraud seem to believe that the self-correction of false ideas is so effective that science can live with the occasional case of data fabrication and manipulation as long as some incremental improvements are made. For example, Marcia McNutt (2020), former editor of *Science*, writes: "The research community has built up a healthy degree of skepticism concerning the results of any single study unless verified by another research group, preferentially using alternative investigation methods and generalized to other systems. A single study is merely a suggestion of a promising direction that is worth verifying. . . . Funder mandates and journal policies to promote openness and transparency are certainly an improvement in creating an environment that could allow the self-correction tradition of science to proceed in a more efficient and orderly way." Similarly, Lee McIntyre (2019) believes that, over time, the wisdom of crowds ensures that scientists collectively succeed in identifying and correcting wrong ideas that are published, whether due to honest error or fraudulent research practices.

Scientific knowledge develops systematically, and over time science helps mankind to understand the world better than before. Scientific progress is real. Moreover, scientific debates are usually won by the camp that makes the better predictions. For example, in the late 1960s and early 1970s, the emerging field of climate science was divided into two camps. One group predicted global cooling because rising aerosol emissions allow fewer solar rays to reach the Earth's surface, and the resulting global dimming could "reduce the surface temperature by as much as 3.5°K. If sustained over a period of several years, such a temperature decrease over the whole globe is believed to be sufficient to trigger an ice age" (Rasool and Schneider 1971: 138). The other group expected, on the contrary, that the Earth's temperature will rise due to a greenhouse effect caused by ever higher CO_2 emissions. Despite the lack of enough reliable empirical evidence at the time, the two camps grew at very different rates, so that by the end of the 1970s the number of articles predicting global warming vastly outnumbered those predicting global cooling.

Others see things differently. John Ioannidis, the skeptic who suggested that most published empirical results are wrong, questions the effectiveness of science's ability to self-correct and to do so in a timely manner: "The self-correction principle does not mean that all science is correct and credible. . . . Even if we believe that properly conducted science will asymptotically trend towards perfect credibility, there is no guarantee that scientific credibility continuously improves and that there are no gap periods during which scientific credibility drops or sinks (slightly or dramatically)" (Ioannidis 2012: 645).

Indeed, time does not always fly in science. It took generations of researchers to overcome the idea that the Earth was flat and that the Sun revolved around the Earth, that phlogiston explained fire, and that ether filled the gap between the stars and the planets. More recently, it took *The Lancet* twelve years to retract Wakefield's fraudulent article, while the excommunication of Macchiarini from the scientific community took four years, during which the Karolinska Institute could not decide whether Macchiarini had violated scientific rules. And Francesca Gino experienced a meteoric rise in the field of social psychology between the publication of her articles on dishonesty in the early 2010s and the retraction of four of these articles in 2021.

Science does self-correct. No doubt. But the process of self-correction is slow. In the long run, science may have successfully corrected itself, but as John Maynard Keynes observed, in the long run we are all dead. In the meantime, data fabrication, data manipulation, and the tweaking of results substantively distort our knowledge and the scientific marketplace. Self-correction is not a magic bullet that works overnight (Vazire 2019).

We see at least three major problems with placing too much hope in self-correction as a deterrent:

1. Self-correction is an improbable mechanism. The vast majority of published empirical analyses will never be replicated or included in a meta-analysis. Nor will they arouse sufficient suspicion or interest among semi-professional fraud hunters, unless they attract major media or other attention. This applies not only to articles that make a minor contribution and are rarely cited, but also to highly cited articles. Fraudulent research with results that are well within the received wisdom of existing knowledge is particularly unlikely to be detected.
2. Self-correction is a costly mechanism (Harris 2017: 39). Meta-analyses are often cited, but they are time-consuming. Researchers who conduct replication analyses have little to gain. Replication studies are rarely cited, and journal editors, who strive for high impact factors, are reluctant to publish them.[3] Neither studies whose results have been identified as outliers in meta-analyses nor non-replicable studies are automatically fraudulent. Far from it. Only additional forensic work can turn suspicion into harder evidence, and that is hard work. Semi-professional fraud hunters spend an enormous amount of time and effort identifying fraudsters, which has a significant opportunity cost. The pursuit of fraudsters is therefore a public good that is woefully undersupplied.
3. Self-correction is a slow mechanism. Until replication studies demonstrate the lack of replicability and further scrutiny of each non-replicable paper demonstrates that outright fraud or large-scale tweaking is the cause of non-replicability, much water will flow down the river Rhine, as the German saying goes. Meanwhile, fraudsters and tweakers will do very well in the academic market for jobs, promotions, attention, fame, and research grants.

Whistleblowing

A major problem with relying on science to self-correct or research institutions to sanction fraudsters is the low probability of detection.

Whistleblowing is a potentially powerful tool in this regard. Many detected fraudsters have been brought down by whistleblowers. Whistleblowers provide an important public good to the scientific community and, along with semi-professional fraud hunters, are the most effective weapon in science's arsenal. And yet whistleblowers are not popular. The accused, the accused's colleagues and allies, and some outsiders may suspect whistleblowers of envying the success of other scientists, of fouling their own nests, of spreading mistrust, of undermining a sense of collegiality, and of unfairly damaging the reputation of scientists.

Whistleblowers often receive little sympathy and support from their institutions. University leaders seem to think that their first duty is to protect their scientific staff from accusations until incontrovertible evidence of their fraud has been gathered against them. According to Lubalin and Matheson (1999), between 15 and 43 percent of all whistleblowers experience some form of pressure to drop their allegations. Retaliation includes attempts to destroy the whistleblower's scientific reputation (Rothschild 2013), reduce or revoke their financial support (Rhodes and Strain 2004), prevent a renewal of their contracts or trigger a denial of tenure (Malek 2010), isolate the whistleblower (Kuhar 2009), question their credibility and judgment (Vie 2020), or describe them as mentally unstable (Kenny et al. 2019). Given this state of affairs, it takes courage to come forward and blow the whistle.

Robert Sprague, one of the first whistleblowers in science, is such a man.[4] According to *The Chicago Tribune*, he is a "hero," the man who "broke science's brotherhood of silence."[5] He blew the whistle on Stephen Breuning, who, in those days, was famous, to say the least. According to a count of Goldfarb and Pritchard's "Ethics in the Science Classroom" online guide,[6] between 1979 and 1983 Breuning contributed to "34% of all published research on the psychopharmacology of mentally retarded people." After earning a PhD in psychology from the Illinois Institute of Technology in 1977, he accepted an offer from the Western Psychiatric Institute & Clinic at the University of Pittsburgh in 1981.

Two years later, Robert Sprague, a colleague working with Breuning on a grant proposal, saw major irregularities in research reports and suspected that Breuning fabricated results on tranquilizer dosages for children with mental disabilities. Sprague reported the apparent data errors to the National Institute of Mental Health, along with detailed documentation of the research fraud. Breuning quickly admitted to fabricating data "concerning the effects of psychotropic medication" on those with mental disabilities.[7] He admitted that he fabricated data to win a $160,000 research grant from the National Institute of Mental Health. Five years later, Breuning became one of the first scientists to be brought to trial. He pleaded guilty.

The story sounds straightforward, but for Sprague it was no walk in the park. The University of Pittsburgh needed four investigations to conclude that scientific misconduct had occurred. The final verdict came fourteen months after Breuning had pleaded guilty and resigned. Sprague, who suspected a cover-up, clashed repeatedly with university leaders. George A. Huber, vice president of the university's medical and health care division, wrote a letter to Sprague saying that if he repeated the allegations of a cover-up, "we will take whatever lawful corrective measures are necessary and appropriate, including the filing of legal action."[8] Wow. George A. Huber retired in 2007.

Whistleblowing cases are complicated because, in the absence of incontrovertible evidence, the allegations have to be investigated. Of course, they may turn out to be unfounded, either because they are maliciously invented or, more likely, because there is insufficient evidence that anything more than sloppiness was involved. Matters get particularly complicated when whistleblowers do not stick to the facts of their allegations but engage in ad hominem attacks that discredit the entire whistleblowing case in the eyes of many outsiders, regardless of the merits of the allegations.

The case of whistleblower Markus Kühbacher illustrates this well. His story has twists and turns. The short version goes like this: In late 2007, early 2008, Markus Kühbacher and Nikolai Savaskan, both researchers at the Charité in Berlin, had a conflict over data manipulation in a joint manuscript—a manuscript that was never published. The Head of the Institute for Cell Biology and Neurobiology, Center for Anatomy, Professor Nitsch, sided with Savaskan in an attempt to mediate the conflict. Not happy about the outcome, Kühbacher took a closer look at Nitsch's publications and found irregularities. Kühbacher blew the whistle, the conflict spread, and eventually the university began to defend its professor against the whistleblower. The final twist was that the article by Nitsch, Savaskan, and four other coauthors that Kühbacher had criticized was retracted by the *FASEB* journal because "the publication contains gross flaws" in at least one of the figures published. Retraction Watch's coverage of this complicated story leaves open whether the figure had been manipulated or whether it suffered from honest mistakes.[9] However, a committee installed by the Charité concluded that the article merely contained honest errors that could be corrected. Upon request by the Charité, the journal has in the meantime republished the original article with the addition of an erratum and addendum written by the authors. Retraction Watch calls this a "retracted retraction."[10]

So far, so complicated, but perhaps still typical. Until another blog entry on the Retraction Watch website brought to light that Kühbacher himself was guilty of unsavory ad hominem attacks, whatever the merits of his fraud allegations. This blog published an email sent by the Johannes Gutenberg University in Mainz, Germany, to more than ninety staff members. The letter was signed by the university's president, the vice president for research, and an ombudsperson. It is worth quoting the email in detail:

> In the recent past, a certain Dr. Markus Kühbacher has repeatedly raised allegations of scientific misconduct against a colleague of our University, Professor Robert Nitsch. These allegations concerned falsification and fabrication of data. Via e-mail, Dr. Kühbacher brought this to the attention of numerous scientists, scientific journals, government offices, funding organizations, and the press. . . . The committee of the Johannes Gutenberg University in charge of the investigation of scientific misconduct performed several investigations into publications by Professor Nitsch in high-ranking scientific journals. In addition, we are aware of inquiries performed by several of these journals themselves. In all cases that came to our attention, the investigations produced no evidence for any involvement of Professor Nitsch in any kind of scientific misconduct. Also in the future, the Johannes Gutenberg University will take any allegation of scientific dishonesty seriously and its designated committee will thoroughly investigate any such accusation. We are convinced that scientific misconduct poses a significant threat to science and society. We would like to inform you, however, that we received hundreds of e-mails form Dr. Kühbacher in the meantime. Many of those e-mails lack scientific standards and contain personal insults to Professor Nitsch and his family.

The statement goes on to quote several of these emails from Kühbacher, which revolve around the sexual infidelity of Nitsch's former wife, a highly unusual move for a university.

Whatever the merits of Kühbacher's allegations of scientific fraud, his ad hominem attacks suggest a personal agenda. As the lively reader exchange on the blog post suggests, this does not detract from his whistleblowing case in the eyes of some but discredits it in the eyes of others. At the very least, his whistleblowing case has been weakened because of the widespread suspicion that whistleblowers are pursuing personal vendettas rather than protecting scientific honesty.

One case, various victims: Savaskan, Nitsch, and Kühbacher. And then there is another victim: whistleblowing in science. Whistleblowing can turn from public good into public bad if the parties involved do not follow a clearly defined protocol but let their interactions morph into public mudslinging.

Of course, the Kühbacher case is exceptional, a true mess. However, there are more cases, in fact many more cases, in which whistleblowing became extraordinarily costly for the whistleblower. Karolinska protected Macchiarini and, as Retraction Watch puts it, "whistleblowers paid the price."[11] And this is only one example: The number of whistleblowers who have been fired or otherwise punished by their universities is growing rapidly. And in the United States, Eugenie Reich has started a law firm that provides legal advice to whistleblowers and tries to hold research institutions accountable when they try to cover up scientific misconduct.[12]

What constitutes "best practice" in handling whistleblowing will vary from context to context, but it has five essential elements:

- The parties do not release any information to the public, to the staff of the institution concerned, or to the media before a decision has been taken.
- The institution concerned remains neutral.
- An independent committee is set up as soon as possible to investigate the allegations and ultimately make a recommendation or, better still, a decision.
- The recommendation or decision is accepted and respected by all parties.
- On the basis of the report, the institution informs the relevant journals of the case, the conclusion of the committee, and the final decision.

This process, if well organized, protects the rights of both the whistleblower and the accused. Yet whistleblowing remains underutilized. If this is due to the fact that whistleblowers have little to gain and much to lose, the question is how research institutions can improve the benefit-to-cost ratio of whistleblowing. At present, they are driven more by a sense of moral justice than by material interests. Proposals to further encourage whistleblowing, possibly through financial incentives, are easy to make but difficult to design and implement. Greater incentives would probably lead to more correct and more false allegations. Whether this would be an improvement is unclear and remains controversial.

In any case, encouraging whistleblowing through financial or other incentives will never completely solve the problem. Whistleblowing may work against fraud in some cases, but not where there are no witnesses or where witnesses are unwilling to come forward. It is toothless against tweaking. To the best of our knowledge, no whistleblower has ever made an allegation about "mere" tweaking. In fact, such a case is unlikely to end well for the whistleblower and their case. The investigators would likely conclude that evidence remains insufficient. The Wansinks of the world who have left blatant evidence of tweaking in the public domain are the exception, not the rule.

Does that mean that whistleblowing has failed? As with the self-correction of science, whistleblowing usually works in rare cases but remains an unreliable instrument against scientific fraud across the board. Whistleblowing faces shortcomings similar to the self-correction of science as a mechanism to deter research fraud:

1. Whistleblowing is a mechanism unlikely to occur. Whistleblowers must have intimate knowledge of scientific misconduct and the courage to come forward and accuse potentially powerful colleagues who often have the backing of their research institutions. As a result, whistleblowing remains a rare event that is far from systematic.
2. Whistleblowing can be actively rendered unlikely to occur. Fraudsters can use all sorts of strategies to minimize the risk of whistleblowing. Doing the research alone works well enough, and if writing single-authored articles is not common practice in their research area, they may invent co-authors to make their manuscripts appear typical.
3. Whistleblowing is a costly mechanism. Unless they can remain anonymous, whistleblowers expose themselves to scrutiny and possible retaliation. Even if they remain anonymous, the psychological toll on them is high as they are questioned by investigative committees and possibly the police. Being a whistleblower is no fun.
4. Whistleblowing is not always slow but it can be. Witnesses need to be convinced that scientific misconduct is systematic and not just a one-off event or mere sloppiness before they will come forward and raise concerns. In most cases, a lot of time and evidence has to be gathered before someone blows the whistle.

Sanctions

If self-correction of science and whistleblowing are improbable, costly, and slow mechanisms, other measures are needed to deter fraud. Potential fraudsters can be deterred if detection leads to legal or social sanctions. In fact, the lower the likelihood of detection, the more severe sanctions need to be. Deterrence results from both the probability of detection and the expected severity of sanctions.

Unmasked fraudsters often, albeit not always, lose their jobs or are disciplined in less drastic ways, for example, by being barred from receiving research grants. A small but slowly growing number have been jailed. But these are the high-profile cases, the pop stars of scientific fraud, if you like. For less prominent cases, what happens in science stays in science. Cases of data fraud are usually handled by three types of academic institutions: universities and research institutes, academic journals, and funding agencies. Universities and research institutes carry the primary responsibility for preventing, detecting, and sanctioning scientific fraud. These institutions have procedures for investigating alleged cases of scientific misconduct and for enforcing decisions. As these cases are rare, research ethics committees usually establish ad hoc misconduct panels that investigate suspected cases, collect evidence, write reports, and make recommendations. These reports are often, but not always, made public. In some cases, research institutions are prohibited by law from publishing these reports since disciplinary proceedings are confidential to protect employees. Research organizations have a range of sanctions at their disposal, including the sharpest weapon in the repertoire of measures: the termination of the contract with the fraudster. Journals will also set up an ad hoc committee to investigate whether articles they have published are fraudulent and need to be retracted. In the majority of cases, however, journals contact the corresponding author's institution and ask them to investigate. Journals are not well equipped to investigate data fraud and have only minor sanctioning measures available. At most they will retract the fraudulent article or require a correction. Funding agencies can also impose sanctions. They may temporarily or permanently exclude identified fraudsters from further research funding.

Overlapping responsibilities and the existing rules of confidentiality cause problems for deterrence. For example, universities or research institutes that find a scientist guilty of fraud do not always inform the editors of journals that have published this scientist's fraudulent research papers. And if journals are being informed, they may delay retractions or abstain from retracting fraudulent articles. Other fraudsters manage to move to a different university after their previous university has started an investigation of scientific misconduct and, strangely, get away.

The case of Jatinder Ahluwalia exemplifies the problems universities have in dealing with highly mobile fraudulent researchers. Ahluwalia's meandering career began as a PhD student at Cambridge University. He was formally dismissed from Cambridge on February 18, 1998, because his supervisor found that he had falsified evidence and refused to continue his supervision.[13] One year later, Ahluwalia continued his studies at Imperial College London, where he kept his dismissal from Cambridge to himself. Imperial awarded him with a PhD in March 2003. In 2004, he published an article on innate immunity in *Nature*. In the meantime, Ahluwalia had commenced a postdoc position at University College London (UCL) where a misconduct investigation was triggered by Anthony Segal, who had co-authored Ahluwalia's *Nature* article. In 2008 Segal publicly accused Ahluwalia of having committed multiple acts of research misconduct. UCL installed a screening panel, which in 2010 concluded that Ahluwalia had produced files with data recordings that did not match the original computer files.

The UCL ethics committee later decided that "Dr Ahluwalia deliberately renumbered the files to deceive Professor Clapp as to the results of his patch clamping experiments. The fact that incorrect file names only occurred in the note given to Professor Clapp and did not occur elsewhere in Dr Ahluwalia's records suggests that this renaming was a deliberate act rather than an accidental error."[14] Segal wrote to *Nature* asking to retract the publications shortly after the UCL report was published (Ahluwalia refused to sign the letter).[15] In October 2007, Ahluwalia moved on from UCL to the University of East London, where he continued to teach and supervise PhD students. He published three single-authored articles between 2008 and 2010. In July 2011, eight months after the UCL report was published, the University of East London announced that Ahluwalia no longer worked at their institution. One month later, in August 2011, Imperial College started an investigation into Ahluwalia's PhD, the retracted *Nature* article was a cornerstone of his dissertation, but found "no evidence that Ahluwalia had committed fraud or misconduct in his work at Imperial."[16]

This case amply demonstrates that investigations into scientific fraud are impeded when multiple research institutions are involved and when researchers repeatedly change their employers. Each involved university established its own investigation panel, potentially doubling the work of other committees, and each committee had access to only partial information. If evidence can be found at one university, the same evidence does not necessarily occur elsewhere, at other universities. It is even less likely that committees share the information they have. To make matters worse, recommendations and decisions have been at odds with each other. Finally, as Imperial College noted, the work of the evaluation committee was hampered by the lackluster cooperation of the industrial collaborator (Novartis) on evaluating the scientific integrity, or lack thereof, of Ahluwalia's dissertation project.

The Ahluwalia case seems somewhat atypical in two respects, one fortunate for science, one not so much. The evaluation of the integrity (or absence thereof) of Ahluwalia's research practices took an unduly long time. A large number of institutions were involved, and these did not communicate with each other and, to make matters worse, came to different conclusions. Universities appear reluctant to take responsibility for cases of scientific misconduct. The costs associated with establishing and maintaining committees to handle such cases are substantial, and there is a risk that the reputations of the universities involved will be adversely affected unless the case is handled exceptionally well. Not every university handles cases of potential misconduct as well as Harvard University did in the case of Francesca Gino. At the end of the process, however, Ahluwalia has left academia or he continues to publish under a different name: his most recent publication we were able to find dates back to 2012.

Less fortunately for science, his departure from science cannot necessarily be generalized. Though the scientific careers of those found guilty of data fraud are often over, not all fraudsters actually have to leave academia, at least not in the life sciences. In a study titled "Life After Research Misconduct," Kyle Galbraith, research integrity officer at the University of Illinois in Urbana with a PhD from Vanderbilt, arrives at findings that are rather worrying for science. He investigates the whereabouts of 284 researchers who have been mentioned in the annual reports of the Office of Research Integrity and had engaged in scientific misconduct—fabrication, falsification, or plagiarism—in research funded by the National Institutes of Health. As Galbraith (2017: 26) writes, "for a study like this one, these sources provide sufficient details about the specific misconduct cases." In total, Galbraith's study covers 237 cases of data falsification, 155 cases of data fabrication, and twenty-five cases of plagiarism. Note that researchers can be guilty of multiple forms of research misconduct. Data fraud, thus, dominates the sample.

Galbraith collected information on whether these fraudsters continued to publish in scientific journals and whether they received research funding after their fraud was detected. For roughly half of the fraudsters (47.2 percent or 134 out of 284), Galbraith found evidence of research activities after the detection of fraud. More importantly, Galbraith found that of 81 tenured researchers, at least 54 appear to have continued with research, while only 9 out of 64 individuals with auxiliary positions such as lab technicians were able to continue their careers. The higher up the "food chain" fraudsters are, the more likely they are to get a second chance.[17] The same pattern emerges for research funding: Full professors who committed fraud are at least four times as likely to subsequently receive research funding than graduate students and postdoctoral researchers. Roughly 15 percent of fraudulent academic faculty received research funding *after* their fraudulent papers had been detected.

Do fraudsters deserve a second chance? From a criminal justice perspective, very few fraudsters, perhaps only those who endanger lives, do not deserve a second chance. After all, fraud is not murder and people can change. However, from a deterrence perspective, a second chance further reduces the already small deterrent effect of sanctions. However, it seems far from clear that deterrence would be more effective if no detected fraudulent researcher ever got a second chance, all were fired, and the small number who played it too colorfully ended up in jail for a year or two.

Do sanctioning mechanisms represent a strong enough deterrent? If the goal is to entirely prevent data fraud, then clearly the answer is disappointing. We see at least three major reasons that prevent scientific institutions from solving the problem of data fraud with sanctioning mechanisms:

1. Scientific institutions have the means to sanction fraudulent researchers, but they do not have the means to monitor the behavior of scientists or to evaluate the integrity of a random draw of relevant publications.

Universities do not closely monitor the activities of their faculty; funding agencies do not check whether their funds lead to sound, honest publications; and journals cannot replicate and reanalyze what they publish.

2. Sanctions are costly for the sanctioning institution. No university welcomes when one of its faculty members commits data fraud; no journal editor likes to retract articles they have accepted. As a result, institutions are hesitant and reluctant to be proactive and often need to be nudged to implement policies that prevent data fraud.
3. Sanctions can only come after detection, and the evaluation process is slow. Further complications are to be expected when more than one research institution is involved. In the meantime, fraudsters and tweakers will do very well in the academic market for jobs, promotions, attention, fame, and research grants.

Criminalization

Detected fraudsters often lose their jobs, their sources of income, and their reputations. Their attempts to remain on the fringes of academia fail, albeit with some notable exceptions: For example, though Tessier-Lavigne stepped down as president of Stanford University, he retained his professorship and even became CEO of Xaira Therapeutics, a biotech drug discovery startup that heavily relies on the use of artificial intelligence. Nevertheless, in many cases, social pressure prevents academic institutions from giving detected fraudsters a second chance.

But police and courts usually do not get involved. Richard Smith, former editor of the *British Medical Journal*, complains that while former Volkswagen engineering manager Oliver Schmidt was sentenced to seven years in prison in the United States for his part in manipulating emissions test results, a scientist who "invents data, defrauds funders and publishes fabricated data that may lead to patient harm is highly unlikely to face criminal charges" (Collier 2015: 1273).

It is true that researchers face little risk from the criminal justice system for fabricating and manipulating data. They may suffer reputational damage and lose their jobs, but the vast majority of cases detected are not considered serious enough to result in criminal charges. Of the approximately ninety cases of violations of Canada's Tri-Agency Framework on Responsible Conduct of Research, only three or four are considered serious, and even fewer are considered criminal offenses (Collier 2015: 1273). What makes a case serious is not the extent of the data fraud itself, but the misuse of the research funds received by the fraudulent researcher. Worse, in many countries, manipulating the published results of funded research does not trigger a criminal investigation. Only if the grant application is fraudulent is the misconduct considered a crime.

Like Canada, Germany remains lenient on fraudsters. The criminal code distinguishes between ordinary fraud and especially serious fraud, which is defined as organized and repeated fraud that harms a large number of people. It is doubtful that German courts classify scientific fraud cases as serious. If they do not, the prison sentence is limited to six months. In the Netherlands, Diederik Stapel, one of the few fraudsters to be prosecuted, agreed to perform 120 hours of community service and lose financial benefits equivalent to a year and a half of his former salary. This lenient policy may change in the future, however, because a German court has just ruled that scientific freedom ends when scientific standards are violated in a publication that seeks to gain knowledge of the truth. In other words, if a publication pretends to be scientific, then scientific standards apply.[18] Sounds logical.

Some wish to change the rules of the game. The title "Call the Cops" of an editorial in *Nature* says it all: "Science likes to shelter its crooks with euphemisms. The prefix 'research' softens fraud, and to deliberately obtain public money through deception gets labelled misconduct, among other things. This reflects the fact that the crime is viewed as being against professional standards rather than against the laws of wider society" (Nature 2013). This editorial was seconded (and cited) in November 2016 by Marylin McMahon, law professor at Deakin University. In an article for *The Conversation*, McMahon advocated tougher legal action: "If the risk to reputation and all that follows isn't enough to deter someone from such scientific fraud, then what other steps can science take to maintain the integrity of any published research? . . . The re-framing of some forms of scientific misconduct as criminal fraud recognizes that scientific fraud involving the fabrication of research and/or results in circumstances where private or public funding has been sought or obtained is similar to other forms of fraud."[19] And Richard Smith concludes that all other strategies have failed: "After 30 years of observing how science deals with the problem, I have sadly come to the conclusion that it should be a crime, for three main reasons. First, in a lot of cases, people have been given substantial grants to do honest research, so it really is no different from financial fraud or theft. Second, we have a whole criminal justice system involved that is in the business of gathering and weighing evidence, which universities and other employers of researchers are not very good at. And finally, science itself has failed to deal adequately with research misconduct."[20]

The strongest case for the criminalization of scientific fraud has been made by Chris Said, a data scientist at Propel Inc. Said argues that scientific fraud can be exceptionally costly, using as an example the case of Sylvain Lesné, who in 2006 published an article in *Nature* claiming that amyloid oligomers could cause Alzheimer's disease. The article is highly cited and, according to Said, has contributed to "spur up to $287 million of research into the oligomer hypothesis."[21]

In June 2024, this article was retracted by the authors, with all authors except Lesné agreeing to the retraction, stating that "concerns have been raised regarding figures in this article, including Fig. 2c and Supplementary Fig. 4, which show signs of excessive manipulation, including splicing, duplication and the use of an eraser tool. The data cannot be verified from the records" (Lesné et al. 2024). Moreover, the oligomer hypothesis has failed in every clinical trial that has been attempted. But, of course, an incorrect scientific prediction does not prove that data have been manipulated.

Said's argument for criminalizing scientific fraud, however, does not depend on the economic costs of data fraud. He also points out that if we assume that the oligomer hypothesis delayed the successful treatment of Alzheimer's disease by just one year, 36 million quality-adjusted life years would have been lost—a figure that, according to Said, exceeds the American loss of quality-adjusted life years during World War II.

But the truly dramatic effect he estimates rests entirely on the assumption that one single fraudulent article delayed successful treatment of Alzheimer's by one year. Yet we do not know what the counterfactual outcome would have been had Lesné's article not been published. The oligomer hypothesis has been around since the late 1990s and has been advocated by many scientists, including perfectly honest researchers. Sure, Lesné's article may have led much research and research money astray. But it seems very far-fetched to assume that this article delayed successful Alzheimer's treatment, which to date does not exist, by one year and thus caused the loss of 36 million quality-adjusted life years. Moreover, Said's argument rests solely on his interpretation of an extreme case that is not representative of scientific fraud and its consequences. Should a Dutch court have sent Stapel to prison even though his fraudulent research probably did not kill anyone? Is it a lesser crime to manipulate social science research than medical research? Perhaps this makes perfect sense, but then criminalization depends on its consequences rather than the scientific fraud itself.

Despite these strong advocates for the criminalization of scientific fraud, criminal justice systems have been involved in only a few cases of scientific fraud in just a handful of countries. Dong-Pyou Han was sentenced to almost five years in prison for manipulating the results of HIV vaccine analyses. Steven Eaton was sentenced to three months in prison under the UK's 1999 Good Laboratory Practice Regulations. In 2006, Eric Poehlman became the first US scientist to be jailed for making false statements on a grant application. Poehlman was sentenced to one year and one day in prison, fined $180,000, and banned from receiving federal research funding for life. The fact that the whistleblower received about 15 percent of the fine is in line with US law, but unusual for a data fraud offense outside the United States. Scott Reuben was sentenced to six months in prison. Reuben worked on painkillers and was considered an expert in "post-operative multimodal analgesia therapy," a treatment strategy that combines different drugs other than opioids. In 2005, he received a research grant from Pfizer. For $74,000, he promised to test the effect of Celebrex (celecoxib) as part of a multimodal therapy. According to Reuben, he enrolled 200 participants, 100 in the treatment group that received Pfizer's drug and 100 in a control group that received a placebo. However, he never carried out the promised randomized controlled trial. The study Reuben published in *Anesthesia and Analgesia* reported results from falsified data.[22] He pleaded guilty.[23] In Australia in 2012, a whistleblower accused Professors Bruce Murdoch and Caroline Barwood, both at the University of Queensland, of scientific misconduct in relation to an article on Parkinson's disease published in the *European Journal of Neurology*. An investigation found numerous irregularities and no evidence that the research had ever been carried out. What brought Murdoch down was not that he had fabricated the data, but that he had received funding to carry out an experiment but had never done so. The corrupt misuse of research funds was one of the main reasons for Murdoch's conviction. In the end, the university returned two scientific instruments purchased with Murdoch's research funds and AUS$175,000 of an AUS$300,000 research grant awarded to Barwood.[24] Both Murdoch and Barwood were prosecuted by the Australian courts and received two-year suspended jail sentences. Prosecutor Caroline Marco reminded everyone that more than just scientific progress was at stake, saying that Barwood's actions had deprived other researchers of much-needed funding, wasted her university's money, and given false hope to Parkinson's patients.[25]

These cases have in common that they all can be characterized as applied research: The authors promise a solution to a pressing (health) problem. Ranjit Chandra's data fabrication case also ended up in court, but for different reasons. In 2001, Chandra published an article in the journal *Nutrition* claiming that a multivitamin mineral supplement potentially reverses memory problems in the elderly. The same manuscript had previously been rejected by the *British Medical Journal* because reviewers said that the study had "all the hallmarks of complete fabrication."[26] In fact, the article was a fabricated marketing exercise for a product licensed to Javaan Corporation, a company founded by Chandra's daughter, Amrita Chandra Gagnon. While *Nutrition* eventually retracted the article in 2005 and apologized for publishing the manuscript, Memorial University, the university to which Chandra had been affiliated and where the study had allegedly been conducted, refused to investigate the case because Chandra had retired and left the country, claiming that the raw data had been lost in an office move. More criminal energy went into the case. In 2002, an author named Amrit L. Jain published a study in the journal *Nutrition Research* that confirmed Chandra's published findings. Jain claimed to be affiliated with the Medical Clinic and Nursing Home in Jaipur, an institution that did not exist, and used a rented post office box in Canada. Jain's paper was retracted in 2016.

The Chandra case led to two lawsuits: Marilyn Harvey, the whistleblower who had worked with Chandra and brought the scandal to light, sued the Memorial University of Newfoundland (MUN) because "the lack of action by MUN against Chandra caused people to erroneously believe that her allegations against Chandra were unfounded."[27] And in 2015, Chandra filed a lawsuit against the Canadian Broadcasting Corporation (CBC) for airing a film titled "The Secret Life of Dr. Chandra," a film that can still be found on YouTube and that makes a strong case that Chandra fabricated scientific data to enrich himself.[28] The case went to the jury, which found that the content of the CBC broadcast was true. The court ordered Chandra to pay

CAN$1.6 million to cover the CBC's legal costs.[29] No criminal charges were ever brought against Chandra.

Francesca Gino seems to have learned from Chandra and walks in his boots. Suing Harvard University as well as Uri Simonsohn, Joseph P. Simmons, and Leif D. Nelson, she seems to be fighting to regain her reputation or at least to make a living out of a scandal, just as Chandra did.

How likely is criminalization to deter fraud? Sanctions imposed on fraudsters vary in severity. Advocates of criminalizing scientific misconduct often argue that only the threat of severe sanctions deters fraudsters. But the effect seems to remain small. Estimating the deterrent effect of punishment on crime is one of the most difficult research questions, even in areas where the effect is larger and thus easier to detect. Does the death penalty have a deterrent effect on murder and other capital crimes? The empirical evidence remains inconclusive, partly because of methodological problems and partly because estimates vary widely. John Donohue and Justin Wolfers (2006b) reviewed the scholarship on the death penalty, including US time series data, comparisons between the United States and Canada or between US states with and without the death penalty, US state panel data since 1934, a variety of instrumental variable approaches, and two state-specific execution moratoria. They conclude that the results of none of these studies support the hypothesis that the death penalty has a large effect on capital crime. At best, the effect seems to be smaller than the uncertainty in the data: Apparently random fluctuations in homicide rates over time are substantial and arguably larger than the systematic differences between jurisdictions with and without the death penalty.

But deterrence is not the only reason for punishment. Others include the restoration of a sense of moral justice and the reparation of the harm done to victims. Occasionally, victims can be made whole again through compensation and other measures. Criminalizing scientific fraud seems to provide some moral justice by punishing the fraudsters. However, compensation for the victims has always been and remains an unlikely outcome. By the time a fraudster has been brought to justice, the damage has long been done and a diffuse group of people have been harmed. The damage caused by fraud is not limited to the distortion of the truth or the misuse of research funds. Scientific fraud is perhaps primarily about the misallocation of scarce resources among scientists. For every fraudulent researcher who has received a job offer, a promotion, or a research grant, there is an honest researcher who has not. Fraud not only manipulates empirical results; fraud also manipulates the job market and the competition between researchers for honor and glory. Fabrication, manipulation, and tweaking turn fair competition into unfair competition. No one who has lost out in scientific competition because of a competitor's scientific fraud is made whole again by the criminal conviction of the fraudster.

Criminalization is not a promising strategy for deterring scientific fraud, for a range of reasons:

1. Few cases have been prosecuted. According to the Retraction Watch database, more than 5,000 articles were retracted every year since 2021. It is unrealistic to assume that the legal system has the capacity and interest to deal with even a small proportion of these cases.
2. Bringing criminal charges usually requires the identification of an aggrieved party, usually research funding bodies. However, fraud generates more than just one victim, and funding agencies are not the main ones.
3. Criminalization is as slow as whistleblowing and self-correction in science, probably slower. Criminalization requires hard evidence and therefore typically depends on a whistleblower coming forward, police investigating, prosecutors bringing charges, and courts convicting. The few cases that have gone to court have usually resulted in convictions. But these cases are extremely rare.
4. It remains questionable whether it is in the interest of society that more cases of scientific fraud are prosecuted. Of course, some fraudsters would perhaps effectively be deterred from fabricating and manipulating data. This does not mean, however, that these deterred fraudsters now write honest research papers. Since the incentives to manipulate results have not changed, fraudsters may simply switch to manipulations that cannot be criminalized, that is, tweaks. Dal-Re et al. (2020: 8) believe that criminalization backfires: Criminalizing data fabrication and manipulation makes other types of research misconduct, for example, tweaks seem less severe ("a lesser sin") and thus renders these seemingly minor fraudulent techniques more likely rather than less.
5. Whether criminalizing data fabrication and manipulation will significantly reduce the temptation to cheat remains to be seen, but courts are certainly not the answer to tweaks. No one wishes to criminalize tweaks. Courts may be able to figure out whether a study is based on

actually existing or merely fabricated data and with some effort and in rare cases even whether data points have been manipulated, but they are not trained to statistically detect tweaks. Replication studies and robustness tests can raise initial suspicions of tweaking. But lack of replicability and robustness will never convince a judge or jury to send the accused to jail. And rightly so.

Does Deterrence Work?

Science can self-correct, but the process of self-correction usually takes a long time, and the current crisis of a never-ending stream of retracted papers demonstrates that one cannot rely on this mechanism. Whistleblowers play an important role in outing fraudsters, and detected fraudsters are usually sanctioned and punished by the scientific community, research institutions, and, on rare occasions, the law. This does not mean that deterrence works. Just as there is no robust correlation between the death penalty and the number of murders, there is no stable correlation between the number of fraudsters who are detected and punished in one way or another and the number of fraud cases committed.

Increasing the criminalization of fraud would have very little effect. If we doubled, tripled, or quadrupled the number of cases prosecuted, the proportion of unsanctioned cases of scientific fraud would still be very close to 100 percent. What the loss of a job and the ruin of a reputation through public shame cannot achieve, the threat of criminal sanction is unlikely to achieve either. Greater reliance on criminal prosecution encourages potential fraudsters to resort to manipulative techniques that can be easily concealed: the manipulation of empirical results that is observationally indistinguishable from honest specification searches. But this result is hardly desirable, since it does not reduce the manipulation of empirical results, but only the likelihood of detection.

Criminalization of scientific misconduct may well be appropriate in serious cases, when human beings are harmed as in the biomedical sciences. However, criminalization does not solve the problems the scientific community faces: overcoming the replication crisis, effectively discouraging the manipulation of empirical results, and preserving or restoring trust in science. Ironically, science's best hope for getting a grip on the problem of the manipulation of empirical results lies with those who claim that science is broken, researchers like John Ioannidis, Brian A. Nosek, and Uri Simonsohn. It is those who bring the replication crisis to light who at the same time try to develop ways for dealing with these problems, including scientific misconduct. The crisis of science generates the attention that may be needed to overcome it (Jamieson 2018).

It is true that some potential fraudsters can be deterred, but detecting and punishing them is unlikely, slow, and costly. It is impossible to know how many published articles are fraudulent and also impossible to know how many scientists would not have manipulated their empirical results if the scientific community had invested more money, time, and effort into deterring potential fraudsters.

9 Prevention Plans

> Bad people will behave badly, and good people will behave well. This line of argument is overly simplistic.
>
> —OECD Global Science Forum

The Ideal Crime

The ideal crime is one that is never committed. However, it is not always possible to prevent criminal acts. Crime prevention relies on the credible threat of sanctions, with sanctions ranging from merely symbolic punishments to torture and capital punishment. However, even the most draconian sanctions are unable to completely eliminate criminal activity.

In cases where sanctions and criminalization prove ineffective, an alternative approach is to legalize activities that were previously prohibited. For example, in cases where anti-doping policies prove ineffective, legalizing certain forms of doping may be a viable strategy. This logic might also apply to scientific publishing, were it not for the fact that false scientific claims made by a fraudulent scientist can kill. It is true that legalization creates a level playing field between competitors who play by the rules and those who do not. But so would the legalization of murder.

A second proposition suggests the implementation of enhanced training and professional education in research ethics for researchers. Nevertheless, there is no lack of awareness among those who engage in fraudulent practices. Fraudsters are usually aware of their actions, even if they try to ignore this fact, and they do not let integrity stand in the way of another publication.

A third alternative strategy is to eliminate the imperative to publish in academia. In the most extreme case, universities and research institutions could remove the publication record of candidates as a criterion for hiring, tenure, and promotion decisions. The argument posits that the absence of a reward for scholarly achievement would preclude the manipulation of empirical results to imitate scholarly achievement. And then we would increase the teaching load of scientists and call them schoolteachers . . .

Perhaps fortunately for scientific progress, the implementation of these and similar radical reforms of the traditional academic publication process seems highly unlikely. Abandoning science because of some rotten apples seems a bit extreme, like throwing the baby out with the bathwater. As an alternative, some critics have suggested changing the way scientific publication is organized, for example, by eliminating statistical significance as an inferential criterion or by reducing the p-value threshold that divides publishable from non-publishable manuscripts, or by eliminating the peer review process. It is a hard break with the traditions of science, but it is conceivable that science will abolish the traditional peer review process, end academic publishing as we know it, and replace academic journals with paper repositories. As Brembs et al. (2023: 9) argue: "The time is ripe to replace traditional and outdated academic journals with an open, interoperable and community-governed modern scholarly information infrastructure." Academic publishing is a market. If paper repositories and alternative forms of academic publishing were superior, then the likes of *PLOS One* and arXiv would come to dominate academic publishing, and publications in journals with high impact factors would no longer be a door-opener to reputation, fame, and research grants. But this has happened at best in small academic niches. Of course, bad habits die slowly. But unless alternative forms of publishing provide superior benefits to authors, it is difficult to picture academic publishing as we know it ceasing to exist. It is even harder to see how these alternative forms of academic publishing would remove the incentives for scientific fraud.

All of these ideas seem unlikely to be implemented. One idea has already made it into publication policy, and that is the preregistration of research designs, especially experimental designs. Preregistration would probably not prevent people like Diederik Stapel from manipulating data points if it helps them get published. Other fraudsters may not fabricate or manipulate data, but instead perform the experiment first and then preregister their design if the results seem publishable. None of these strategies, even in combination, will completely eliminate data fraud.

Education and Training

Could training in research ethics and a more rigorous curriculum for junior researchers help reduce the incidence of fraud in science? Over the past few years, research ethics courses have become a standard part of PhD programs. While they might not be everyone's favorite, they're unlikely to cause any serious harm.

But are these policies effective? Such an argument, if we want to accept it as an argument, suffers from several shortcomings. Does anyone really believe that Diederik Stapel would not have falsified data if an ethics course teacher had told him two decades earlier that this was wrong and constituted scientific misconduct? Is there any doubt that fraudsters know what they are doing? At the very least, training should be compulsory not just for young scientists but for everyone. The most spectacular cases of serial scientific fraud, with some notable exceptions, have been committed by senior scientists. This is not surprising: Junior scientists have little chance of making the Retraction Watch leaderboard, and it raises suspicion when junior scholars in their early thirties have already published fifty articles.

Perhaps educating scientists about tweaks and why tweaking is wrong might help a little, but not much. In principle, researchers of all ages know deep down that fishing for significant results, hypothesizing after results are known, or other tweaking strategies do not constitute honest research practice. Through education and constant reminders, the unconscious can be transformed into conscious knowledge that may lead some back to the right path. But even if researchers accept that tweaking is dishonest, the incentives to improve the publishability of a manuscript and the incentives to manipulate the results of a paper for ideological or other self-serving reasons remain exactly the same.

A problem that can be solved with a few good words and a bit of education isn't much of a problem at all.

The Scientific Marketplace

A more serious prevention strategy is to remove the pressure to publish in the academic marketplace. The bottleneck in the academic labor market between the PhD and the tenure-track position facilitates scientific fraud, it seems. Universities produce more PhDs than there are open academic or research positions. This bottleneck intensifies competition between junior researchers, and, as professional cycling has shown, fierce competition in a "winner takes all" situation encourages the use of fraudulent means. Of course, there are other bottlenecks in the scientific marketplace, and these certainly exert competitive publication and other pressures on scientists. Highly ranked journals can publish only so many articles before their reputation and standing suffer. Research funding remains limited. As a consequence, only a few research stars can exist in any field.

The survival of junior researchers in the long-distance race for tenure and later full professorship is fought primarily on the basis of publication success. Junior researchers are expected to be productive and to publish. Some scientific disciplines require ten or more publications for tenure. Depending on the tenure-granting university, some, if not the majority, of these articles must be published in prestigious journals, often with a rejection rate of over 90 percent of submitted manuscripts. This is no walk in the park. Perhaps pressures can be reduced. But this strategy compromises the scientific quality of research. Universities that do not recruit the most promising and successful researchers need other selection criteria, criteria such as a strong focus on the best teachers. This strategy, however, does not remove bottlenecks or competitive pressures. It simply changes the criteria, replacing one competitive pressure with another. At the same time, these selection criteria damage the pursuit of excellence in research. If research excellence is no longer rewarded, then academics who have not yet reached the peak of their academic careers invest less in good research. Ironically, a focus on teaching can actually harm the quality of teaching. Good researchers are often better teachers, and removing the demand for good research in the labor market will reduce the supply of both good researchers and good teachers.

Another strategy is to change the nature and criteria for assessing research excellence. The most visible initiative in this respect is the San Francisco Declaration on Research Assessment (DORA). On the surface, the initiative states what everyone can agree upon: The content of an article is more important than the impact factor of the journal in which it is published. No doubt a brilliant article in the *Lower Antarctic Journal of Research*, a journal that is not ranked because it does not exist, is better than some rubbish published in *Nature* or *Science*.

No one denies that a journal's impact factor provides only limited information about individual articles. All that the impact factor conveys is that the average article published in journals with high impact factors is cited more often than the average article published in journals with low impact factors. The impact factor does not say that each article published in the former journal has been cited more often than the most cited article in the journal with the low impact factor. However, DORA does not suggest looking at the citations of an article in relation to those of comparable articles, which is a simple and straightforward way of evaluating the comparative impact of articles. Instead, DORA claims that "the scientific content of a paper is much more important than its publication metrics" and thus urges academic deciders to "consider the impact on policy and practice." DORA suggests that the work of Theodore Maiman is more important than the work of Albert Einstein on the stimulated emission of radiation. After all, Maiman produced the first laser; Einstein just laid the foundations. DORA's claim may occasionally entail some truth: many highly cited articles are useless. However, replacing objective and observable criteria derived from citation metrics with subjective and unobservable criteria is unlikely to achieve the desired result. And it will not prevent data fabrication and manipulation.

Weed (1998: 126) sees little hope for a significant impact from reducing competitive pressures: "It is not clear how reducing—much less eliminating—these factors would reduce scientific misconduct without also reducing some of that which makes science the rigorous and productive enterprise it has become. Perhaps it is a matter of degree. Indeed, to reduce publication pressure, suggested interventions involve emphasizing quality over quantity in academic appointments and promotions as well as eliminating honorary authorship. Although these are reasonable proposals, they may have little impact on the publication pressure inherent in science." We agree.

Even if the competitive pressure could be reduced, this is not the only reason for manipulating empirical results: greed, as in the case of Ranjit Chandra, the self-proclaimed "father of nutritional immunology"; hubris, as in the case of Diederik Stapel and other notorious fraudsters; and ideology, as in the case of LaCour, Reinhardt and Rogoff, Piketty, and many others. Then there are the vanity rents to be made from a successful, highly cited, and mega-influential publication. No matter how we reform the incentive structure in the academic marketplace and try to make it suitable for Generation Z, Einstein will always be more famous than the rest of us. And that hurts. As a consequence, some scientists will always be tempted to replace a 2 with a 4. Because they can, and because that makes them feel better. Closer to Einstein, closer to the next promotion.

Publication Criteria

Fraudsters and tweakers seek publications, and more of them: publications that are placed in more highly ranked journals that generate more citations and more impact both within and outside the academy. Journals set criteria for what makes a manuscript publishable, as they receive far more manuscripts than they can or want to publish in order to maintain their status as a highly selective journal. But one thing is for sure: Statistically insignificant results have a very hard time getting published. If significant results are a minimum requirement for publishability, does abolishing statistical significance as a publication criterion eliminate data fraud and tweaking?

Some journals and some scientific communities have moved away from p-value signaling. *Basic and Applied Social Psychology*, *Health Economics*, and the *Journal of the American Statistical Association* have published editorials and editorial policies that discourage or outright disallow authors to report asterisks and p-values. In its most consequential form, some prohibit researchers from discussing whether their results are significant. In a softer variant, authors are banned from reporting statistical significance and rather asked to frame the discussion in terms of confidence intervals around point estimates and effect sizes. This movement is not, at least not primarily, driven by a desire to eliminate data fraud and manipulation. Instead, it is driven by more fundamental concerns about the serious flaws and shortcomings of statistical significance as an inferential criterion (Gelman and Stern 2006; Gross 2015; Imbens 2021) and as a distraction from the substantive size of an effect: the quantity of interest that really matters (Gill 1999; Ziliak and McCloskey 2013; Esarey and Danneman 2015).

Statistical significance says nothing about the probability that a hypothesis is correct. The p-value simply calculates the probability of rejecting the null hypothesis if it is correct, if models are perfectly specified, if the errors happen to be perfectly normally distributed, and if the estimator is unbiased. These are a lot of ifs, and they are usually not met in actual research. However, the shortcomings of the p-value as a statistical criterion are of little interest here. Instead, we will focus on whether banning it from the peer-review and publication process is likely to eliminate fraud and manipulation.

Unfortunately, the answer is: most certainly not. Researchers do not just manipulate p-values. Fraudsters manipulate whatever stands between their manuscript and the journal in which they want to publish. Replacing the p-value with a focus on effect size is a good idea for other reasons, but as long as some criteria exist, empirical models will be tweaked. The incentives to tweak are immanent to the academic system. If substantively important effect size becomes the new relevant quantity of interest that every reviewer focuses on, tweakers will shift their attention from pushing the p-value under the ominous 0.05 threshold to making the substantive predicted effect as large as possible. The same is true for any other quantity of interest that might replace the p-value. If editors wanted to, they could banish statistical significance as a relevant criterion for their journal with the stroke of a pen. But they cannot banish fraud with the stroke of a pen.

Others suggest the exact opposite of abandoning statistical significance. Recognizing that the 0.05 threshold, combined with the file drawer effect, means that a large number of published significant results are likely to be false positives, some believe that lowering the threshold for statistical significance from 0.05 to 0.005 will help (Esarey 2017). Obviously, results that are significant at a p-value of less than 0.005 are considerably less likely to be false positives. At the same time, this criterion is much more likely to cause false negatives.[1]

Whether more rigid significance thresholds help against tweaking is unclear. On the one hand, p-hacking at lower thresholds probably becomes more difficult and requires more brute force. On the other hand, lowering the threshold may lead to "more intense p-hacking and selective reporting" (Amrhein and Greenland 2018): Since manuscripts with a relevant p-value between 0.005 and 0.050 will be more likely to be rejected, researchers will be more tempted to tweak these empirical results until they find a model with a p-values of 0.005 or smaller. At the same time, lowering the p-value threshold for statistical significance will also further and more strongly discriminate against studies with a relatively small sample. Researchers will thus resort to very large datasets, even if these are not best suited for their research questions.

Peer Review

Confidence in science could be resurrected if the peer-review process detected all tweaked papers and all falsified and false empirical results, and if what was peer-reviewed for publication were the truth, the whole truth, and nothing but the truth. If peer review were to have these effects, there would be no manipulation of empirical results.

The reality of the peer-review process is far from this ideal. Despite consuming, according to one conservative estimate, a colossal 100 million hours (Aczel et al. 2021), valued at several billions of dollars, peer review remains an inherently flawed process. The review process has never worked well. Good manuscripts are rejected by good journals, while bad manuscripts are accepted, and vice versa for minor scientific journals. Peer review is a selection process that has certain inherent drawbacks: It is decentralized, inefficient, and arbitrary, and slow, sometimes painfully so.

Peer review is decentralized. Many scholars have an influence on the decision to publish. Rather than being decided by a dictatorial editor or a small editorial team who makes all decisions based on their own judgment, two or sometimes more scholars get a chance to have a say on the manuscript. Decentralization at least potentially widens and deepens the pool of knowledge that editors can draw on for deciding the fate of manuscripts. But decentralization also reduces the accountability of editors and their decisions. Editors tend to hide behind anonymous referee reports. Their letters to authors often reveal that they have not read the manuscript and merely have followed the line of least resistance, rejecting papers if one or more of the reviewers is critical, however unjustified that criticism is, or advancing papers if they need to fill the journal or do not want to upset a famous or well-connected submitting author.

Peer review is both inefficient and arbitrary. The former results from the fact that several reviewers read the same manuscript, comment on the same theory, the same empirical analysis, the same contributions. If they come to the same conclusion, this is inefficient. If they come to a different conclusion, the arbitrariness of the process becomes apparent: If peer review were a bullet-proof mechanism for assessing the quality of a manuscript, reviewers' assessments would roughly be identical, independent of the reviewer. Instead, getting a manuscript past reviewers depends to a large extent on the reviewers assigned to review the manuscript.

Peer review is slow. Journal editors have difficulties finding reviewers. The time from submission to decision, often months and months, is determined by a weakest link problem: The slowest reviewer determines when the editor can make an informed decision. Recently, journals have begun to compete with each other by promising submitting authors a short average review time. Of course, reviewers rarely need four or more weeks to review a manuscript and identify its potential weaknesses. Arguably, however, placing busy reviewers under time pressure will negatively affect the quality of their reviews.

All these drawbacks might be a price worth paying if peer review weeded out fabricated and manipulated papers. However, to our knowledge, few if any fraudulent papers have been detected at the review stage. All the papers that have been retracted for fraud, often years after publication, have by definition gone through the peer review process and not been caught. Admittedly, reviewers cannot easily detect fraud in a paper unless they are trained to detect statistical anomalies in published results. However, evidence shows that the average reviewer does a pretty poor job of finding even fairly blatant basic errors. As Mastroianni (2022) reports, some naughty scientists deliberately added unconcealed errors to manuscripts and sent them out for peer review to test whether these errors were noticed. In the three studies he cites, the reviewers only caught between 25 and 30 percent of the major errors. This does not bode well for the ability of peer review to detect data manipulation and fabrication or tweaking.

We have already met Ranjit Chandra. In all likelihood, he fabricated results and then published a replication study under a false name that supported the fabricated evidence. His case shows that the detection of fraud during the academic peer-review process does not necessarily stop the fraudster from publishing the fraudulent manuscript. Chandra's original fake manuscript was rejected by the *British Medical Journal*. One of the reviewers apparently mentioned that the evidence was fabricated. Strangely, this allegation was not enough to stop the fake paper being published. Chandra instead submitted the manuscript to the journal *Nutrition*, where it was accepted and published, only to be retracted four years later. Clearly, in this case, the peer-review process did not prevent the publication of a fraudulent article in another journal, because journal editors did not and still do not share evidence of research fraud with other journal editors.

Can the peer process review be reformed to significantly strengthen its ability to prevent fabrication and tweaking? Training reviewers is hopeless. There are far too many researchers who review manuscripts; their willingness to be trained converges to zero; and training will not have much impact. Paying reviewers according to the quality of their reviews is a solution welcomed by reviewers. If they are being paid, they may pay more attention, but data fabrication and tweaking cannot reliably be detected by reviewers. It thus remains unlikely that more attention would eliminate tweaking. Mastroianni (2022) believes that peer review can actually exacerbate the situation. Reviewers usually ask for different specifications from those contained in the original manuscript being reviewed. Authors often feel compelled to do what the reviewers ask, even if they disagree with them. If these specifications do not produce statistically significant results, authors will be tempted to tweak them. Mastroianni argues in favor of publishing one's articles in open access repositories. These repositories became more popular during the COVID-19 pandemic because researchers wanted to get their findings out quickly without being slowed down by the normal peer-review process. Papers on these repositories are not peer-reviewed, and the vast majority of them will be forgotten, generating little or no attention or citations. A few others may attract significant attention and will be forensically examined by dozens, perhaps hundreds, of peers who will analyze the work and possibly replicate it. Such a policy increases the likelihood of fraud detection. However, the attention and not the absence of peer review drives the detection.

Compulsory Preregistration

Nosek et al. (2018: 2604) proclaim a "preregistration revolution" and justify its necessity as follows: "just as the present culture provides means (reasoning biases and misuse of statistics), motive (publication), and opportunity (no a priori commitments to predictions) for dysfunctional research practices, the culture is shifting to provide means, motive, and opportunity for rigor and robustness of research practices via preregistration." In its classic form, preregistration means that researchers produce an openly accessible and time-stamped plan before carrying out their research. This plan specifies exactly, precisely, and sufficiently, but also succinctly, what they intend to do, so that reviewers, editors, and readers can easily assess whether the preregistration plan has been followed. The plan states the hypothesis or prediction to be tested and describes the research design, including sampling strategies and sample sizes; the measurement and operationalization of dependent, explanatory, and control variables; the specification of the estimation model; the rules for excluding observations; the rules for conducting any subgroup analysis; and anything else that seems to be essential to the analysis. In an ideal world, preregistration separates research design, hypothesis formulation, and model specification from data collection and analysis. As a result, researchers cannot succumb to the temptation to tweak because they have already committed themselves to a course of action. Or at least that is the optimistic idea.

Preregistration has become fashionable, with the number of studies preregistered on platforms such as AsPredicted.org, the American Economic Association Registry, or the Open Science Foundation website rising sharply over the last half decade (Simmons et al. 2021a: 151). Many researchers, especially experimentalists, have embraced preregistration. Preregistering a study lends credibility to research designs and empirical analyses. Some go so far as to suggest that journals should make preregistration mandatory.

But can preregistration prevent fraud, and if so, are the costs associated with preregistration worth paying?

Preregistration cannot weed out the truly rotten apples. Simmons and co-authors make the case that preregistration is most useful for honest, moral, and competent researchers: "In our experience, p-hacking is not merely easy for moral and competent researchers to *do*, but actually hard for moral and competent researchers to *avoid*" (Simmons et al. 2021a: 154). Outright fraudsters, on the other hand, are fraudsters who can and will fabricate data whether they have prespecified a research plan or not. Simmons, Simonsohn, and Nelson agree that preregistration does not help much against such data fraud.

Methodologists often concede that preregistration is limited to experimental research. Indeed, preregistered observational data analyses remain rare, at least in economics (Christensen and Miguel 2018: 959). Since causal models are observational data analyses with a twist, this concession applies to these models as well. In observational data analyses, there needs to be sufficient room for learning in the form of honest specification searches during the data analysis phase. Every observational data analyst has experienced how model specifications can be improved by running econometric tests or by looking for structure in the residuals.

Andrew Gelman, who usually is straight to the point when he suspects that authors have tweaked their models to get the desired results, makes a similar point in a book with Jennifer Hill and Aki Vehtari (2021). For Gelman, regression analyses of observational data inevitably go through four stages, namely, model building, model fitting, understanding model fits, and evaluating results, to return to the model-building stage and go through the cycle of stages all over again (Gelman et al. 2021: 12–13). It is difficult to see how all these specification searches can be captured in a preregistration plan. Simmons et al. (2021a: 160) acknowledge that "at some basic level, pre-registration is not the right tool for the job." "Job" stands for observational data analyses. They suggest that robustness testing is much more appropriate. Preregistration may actually reduce researchers' incentives to explore the robustness of their findings, as Sara Steegen and collaborators (Steegen et al. 2016: 709) point out: "By preregistering a study, all analytical choices—including the arbitrary ones—are made ahead of time, before collecting the data. . . . The choices cannot be made conditional on the (real) data. However, the considered results are still just the results given one choice combination . . . and their robustness across other reasonable choice alternatives remains hidden."

Some preregistration studies competently deal with specification uncertainty by specifying all the statistical tests and other considerations that ultimately determine the final specification of the model. But these are rare exceptions. Though preregistration does not prevent adjustments during the research process. It simply demands from researchers to explain and justify the adjustments. Still, authors of preregistered studies often invest in data mining and harking activities *before* they actually preregister their research design. Only if researchers know what "works" will they register their design. This strategy has been called "parking" (preregistering after results are known) (Baruch 2024). Nothing prevents tweakers from "preregistering" a study, experimental or otherwise, they have already undertaken. If "parking" replaces "harking" (hypothesizing after results are known) and other tweaking strategies, then little has been won. Apparently, preregistration may at best render tweaking more difficult, but clearly does not prevent it. "Parking" is an almost undetectable fraudulent strategy against which preregistration is toothless. Preregistration cannot prevent tweakers from "merely" preregistering after they have undertaken *some* preliminary or exploratory empirical analysis of which research design generates confirmatory and statistically significant results. Simmons et al. (2021b: 178) contend that gaming preregistration analyses is possible but rare. How do they know? We do not know.

Nothing prevents tweakers from running a pre-test of their experiment or their empirical analyses before they "preregister" their study. Even if the results of a pre-test are insignificant, skilled researchers can calculate how many additional participants are needed to make the results significant. Similarly, they can experiment with different levels of treatment in the pretest to observe which level gives the required result, and they can look at the data to identify subgroups of participants who are particularly responsive to the treatment.

Advocates of preregistration counter these concerns in two ways. First, they argue that the incentives to conduct undisclosed preliminary or exploratory analyses to search for publishable results will disappear if editors and reviewers are willing to commit to publishing the results of a prespecified analysis plan, regardless of whether the results ultimately found are significant or not (Nosek and Lakens 2014). Second, supporters argue that preregistration does not prevent preliminary or exploratory analyses and, somewhat paradoxically, actually makes it easier to conduct these analyses because preregistration allows researchers to clearly and honestly state how much "exploration" went into their research design and model specification, and which results were exploratory and therefore more tentative (Simmons et al. 2021a: 154–155).

We find both arguments unconvincing.

First, it is, of course, true that the incentive to manipulate empirical results for no other reason than to get the manuscript published disappears if journals review preregistered research designs and commit to publishing accepted papers regardless of the outcome, even in the case of null results. Publication bias disappears when empirical results no longer determine whether manuscripts are published. This argument lies at the heart of the movement to publish all manuscripts in repositories and to do away with academic publishers in general and academic journals in particular (Brembs et al. 2023). There is no doubt that this solution is technically feasible: University repositories, global repositories organized by discipline, and publishing platforms such as the Public Library of Science (PLoS), a nonprofit organization that has been profitable since 2010, provide the infrastructure for the publication of large numbers of research papers outside traditional journals. Yet the existence of these platforms has not ended the dominance of academic journals. In fact, traditional journal publication has advantages over the new alternative forms of academic publishing: It gives credibility to the articles they publish because the peer-review process, for all its faults and shortcomings, is more selective than that of PLoS, not to mention repositories, which often publish without peer review. For the time being, it seems that journals and commercial academic publishing will not only survive but will continue to dominate the market.

Journals and their editorial boards aim to achieve a high impact factor, which enhances the journal's ranking and attracts more high-quality manuscript submissions. Given this competitive environment, it is unlikely that these journals will continue to publish large numbers of preregistered trials if the proportion of those trials yielding null results increases (Kaplan and Irvin 2015; Papageorgiou et al. 2018; Scheel et al. 2021). Articles with null results often receive minimal attention, resulting in fewer citations. This, in turn, lowers the journal's impact factor, and editors are likely to lose interest in publishing such findings.

Second, and related to the first point, preregistration does not eliminate the incentive to avoid null results. Its proponents fail to recognize that tweakers are not just looking for publications but publications with statistically significant results. Only these will attract attention, provide evidence for the tweaker's preferred hypothesis, and pave the way to attention and fame. Null results and statistically significant results do not have the same benefit for authors. Peer review of preregistered trials with a binding publication decision may eliminate the possibility of manipulating empirical results to increase the chances of getting a manuscript published. However, researchers do not publish just for the sake of seeing their name in a journal; they seek attention, fame, promotions, and research grants, or simply have ideological biases. None of these goals can usually be reached by publishing null results. Publications are not an end, they are a means. And publications of null results do not serve these means, at least not to the same extent as publications with important contributions. As a result, the incentives to manipulate empirical results remain, even if science were to publish all papers regardless of results.

We are therefore unpersuaded by the arguments for preregistration, as they misunderstand the incentives of manipulators. At the same time, our main argument against preregistration revolves around its unintended consequences. The purpose of preregistration is to lend credibility to the results of an empirical analysis. Credibility increases the likelihood of publication. Researchers therefore have an interest in preregistering their research project, all other things being equal. Preregistration gives readers a sense of the credibility of empirical results, whether justified or not.

There are two types of credibility: earned and unearned. Earned credibility is a good thing. But undeserved credibility is a bad thing. As Nosek and Lakens (2014: 138) acknowledge, preregistration cannot prevent false results based on flawed analyses. A preregistered analysis may be poorly conducted or may have the estimation model seriously misspecified, the study design suboptimal, or other issues. In fact, if one accepts that the majority of specification searches and other learning during the data analysis stages are honest and not fraudulent, and if specification searches and learning are successful and improve the research design and model specification, preregistration makes the average model misspecification worse if researchers refrain from adjusting specifications. Preregistration is no free lunch and may, on average, reduce the validity of causal, statistical, and descriptive inferences.

The concern, therefore, is that preregistration signals credibility that is not (always) deserved. Pham and Oh (2021: 173) argue that "preregistration should not be used as a lazy heuristic for research quality." But if preregistration does not guarantee research quality, what good does it do? One possible answer is that preregistration exposes poor research quality and outright fraud. If the preregistered study collects privately held data, fraudsters can always manipulate data points to get the results they want without having to fear detection.

Why do fraudsters preregister an analysis that they can manipulate if necessary? The answer is to gain credibility for their analyses and increase the publishability of their research. For honest researchers, preregistration is often just a nuisance, but for fraudsters, preregistration may well be the dominant strategy to gain undeserved credibility for their analyses.

Research Transparency

The alternative to preregistration is not to do nothing. A culture of greater research transparency and openness, in which journals require and enforce transparency of design and analysis, reporting standards and disclosure requirements, and public publication of all data and code with strict limits on which data can be declared confidential and therefore private, offers an alternative. At the very least, this alternative allows easier replication and extensive robustness testing by others. Instead of having to trust researchers, they will be held accountable and that is an advantage because "transparency is superior to trust" (Munafò et al. 2017: 5).

What Christensen et al. (2019: 175) report about the unavailability of data for published articles is shocking, but not entirely surprising: "Sociologist Cristobal Young had students request replication materials for sociology articles whose analyses they wanted to try to reproduce. Only 15 out of 53 requests (28 per cent) were successful." Political scientist Thomas Carsey conducted a similar exercise and reported that students "often come away from this assignment frustrated, shocked, and rather disappointed" (Carsey 2014). We have had similar experiences with replication requests.

Some journals already strive to follow guidelines to ensure "good scientific practice." Some require authors to declare interests, disclose funding sources, and indicate which co-authors have contributed to which part of the manuscript. Some journals issue guidelines, which they enforce to varying degrees, on how to write up research. Good practice requires authors to describe the study context and design, data collection and description, methods of analysis, reporting of results, study limitations, and availability of materials (Christensen et al. 2019: 148). Researchers may be required to explain and justify in online appendixes all research decisions and choices made and rejected in research design and model specification, as well as paths not taken in empirical research.

Does Prevention Work?

Legalizing academic fraud, eliminating competitive pressures to publish in the academic marketplace, abolishing statistical significance as a de facto minimum requirement for publishability, or instead doubling down on the p-value by requiring a stricter threshold of statistical significance: We have dismissed all these proposals as non-starters. They are either ineffective against fraud and manipulation or have drastic side effects, or both.

Reforms of the academic peer-review process are unlikely to prevent data fabrication and manipulation. Peer review is inherently imperfect and incapable of detecting fraud and manipulation. The high number of articles corrected or retracted for honest mistakes suggests that peer review cannot even reliably eliminate mistakes. However, its imperfections do not render the peer-review process fatally flawed. In Churchill's famous words, democracy is the worst form of government except for all the others. Likewise, peer review represents the worst mechanism for accepting or rejecting manuscripts except for all the others.

We are not sure that the analogy holds. We sympathize with those who suggest that all research should be published without the hurdle of peer review, but instead made available in open access repositories for the wisdom of the crowd to be unleashed upon it. But the peer-reviewed publishing process is tenacious and has survived numerous threats and developments, including the digital revolution and the open access movement, in which authors or their employing institutions pay publishers to make their articles publicly available. As long as good journals exist, scientists will not give up publishing in *Nature, Science, PNAS*, and other highly ranked and respected academic journals, and instead publish their best work in a paper dump called a "repository." Peer-reviewed academic publishing has and will always have problems, but it is not going to die anytime soon.

A growing number of scientists, particularly those working in the experimental tradition, are demanding the preregistration of research plans before any article is considered for publication. However, those who see preregistration as a fail-safe way of preventing tweaking are placing too much faith in its efficacy. To be clear, we are not against preregistration, especially of experimental research designs. However, to believe that preregistration of experimental designs will eliminate tweaking seems hopelessly optimistic. At best it will reduce tweaking, at worst tweakers will hide their dirty strategies behind a veneer of undeserved credibility.

For the analysis of observational data, including in a variety of causal models, preregistration is not very useful, and there are more promising alternatives, namely, requirements to improve the transparency, openness, and reproducibility of published research. Greater transparency and openness in research shift attention to a discussion of research design and model specifications, and away from the oversimplified question of whether results are wrong or not. All results are wrong, but some are close to the truth, while others are outright wrong. Science needs multiple levers to encourage more of the former and prevent more of the latter.

10 Detection Strategies

Denn die einen sind im Dunkeln
und die andern sind im Licht.
Und man sieht nur die im Lichte,
die im Dunkeln sieht man nicht.
—Bertold Brecht

The Invisible and Hidden

If exposed scientific fraudsters were murderers, they would have blood on their hands, the murder weapon in their pockets, and their fingerprints all over the crime scene. In fact, scientific fraudsters seem to get caught only if they invest little in concealing their misdeeds, if they are careless, naive, unskilled, or all of the above.

In many cases of exposed fraudsters, it is not very surprising that their fraud has been discovered. Wansink sent emails describing his fraud. Chandra fabricated implausible data to support equally implausible hypotheses. Stapel was clever enough to fabricate socially desirable research results, but he refused to share raw data with his co-authors. Summerlin, who claimed that he had transplanted tissue from one mouse to another but merely inked the mice, failed to keep his dyed mice out of sight of his lab assistants. LaCour fabricated data suggesting a response rate too good to be true for surveys. Mart Bax not only published about events that never happened, but also on villages that do not exist (Salemink and Verrips 2020). Indeed: many exposed fraudsters behave like bank robbers driving their golden Rolls Royce through town: Here I am, look at me, I am the greatest. Okay, they may not literally drive fancy cars around town, but they publish too much in too little time in too good journals.

Jan Hendrik Schön is one of them. At the age of thirty-two, he published a research paper every eight days; seventeen of these articles appeared in *Nature* and *Science*. He had already won the Otto-Klung-Weberbank Prize for Physics, the Braunschweig Prize, and the Outstanding Young Investigator Award of the Materials Research Society. Apparently, a superstar in the making, and that should have set off a few alarm bells. And then it did: Soon after he published on single-molecule semiconductors, other physicists pointed out that his data showed significant anomalies. Others noted that Schön had duplicated the figures (that he also had fabricated). Publishing fifty papers per year is hard work, and thus any shortcuts were welcome.

Somehow, detection worked in the case of Schön. But science cannot overcome its credibility crisis if only the most incompetent or shameless fraudsters are detected. Science must learn how to detect skilled, competent, and careful manipulators, not to mention those who "only" tweak results rather than fabricate and manipulate data. The holy grail of fraud prevention is an easy-to-use algorithm that identifies fraud and tweaking at the touch of a button. But holy grails are impossible to find.

Forensic Statistic

If results are too nice, they may simply have been manipulated. Fraudsters, especially serial ones, like to take shortcuts and publish evidence that forensic statistical reasoning will show to be extremely unlikely to be genuine and real. When Peter Kranke, Christian Apfel, Norbert Roewer, and M. D. Franke published their critique of the work of Yoshitaka Fujii and co-authors, they used the title "Reported Data on Granisetron and Postoperative Nausea and Vomiting by Fujii et al. Are Incredibly Nice!" (Kranke et al. 2000). Calling someone else's work "incredibly nice" sounds incredibly nice, but the real meaning here is that the results are too good to be true: "With increasing amazement, we noticed that the results reported by Fujii et al. are incredibly nice and we became skeptical when we realized that side effects were almost always identical in all groups" (Kranke et al. 2000: 1004).

In retrospect, it seems surprising that Kranke et al.'s critique of Fujii went nowhere at first. Many more years had to pass by before Steve Yentis, a newly installed editor-in-chief of *Anesthesia*, commissioned a short editorial on Fujii's work, which prompted a flood of responses from the journal's readers. Yentis asked one of these commentators, John Carlisle, to back up his claims with evidence. And so Carlisle took a close look at Fujii's data. After hours and weeks of careful evaluations, he concluded that the odds that these data came from experiments was no more than 10^{-33}, a number not very different from 0. According to Carlisle (2012), he found "unnatural patterns" that "would support the conclusion that these data depart from those that would be expected from random sampling to a sufficient degree that they should not contribute to the evidence base."[1]

Kranke and his co-authors became suspicious of Fujii's work for more than one reason and made their scathing critique public: The reported number of cases seemed implausibly large; the results differed significantly from the results reported by other scientists; and, on closer inspection, they found that "the values of certain parameters and the reported frequencies of side effects were often absolutely identical in different tables. For instance, the side effect 'headache' was identically distributed in independent patient groups."[2] This quote from Peter Kranke comes from a press release from his home institution, the University of Würzburg, celebrating the fact that the editors of twenty-three anesthesiology journals had finally informed the institutions with which Fujii collaborated that almost 200 of his articles should be retracted—twelve years after Kranke et al. (2000) published their "incredibly nice" critique.

Forensic statistics also played a role in the Gino saga. Simonsohn, Simmons, and Nelson used forensic evidence to buttress their allegations of data manipulation. Their case against Gino was strong from the onset because Gino has taken a lazy approach to data manipulation, a strategy that left plenty of suspicious evidence, which was later supported by Harvard's integrity committee, which had, unlike Simonsohn et al., access to some of the raw data. The forensic case against Gino became stronger when Simonsohn and collaborators found compelling traces of scientific fraud in more than one of Gino's articles. A single article with evidence could always be the result of weird data points or an honest mistake. But systematic honest mistakes, mistakes that make results larger and more significant, in more than one article do not appear to be likely.

Two cases, arguably 2:0 for fraud detection through forensic statistical reasoning. But these cases demonstrate the problem with forensic analyses: They are not conclusive, and, more importantly, forensic techniques work best if the fraudster does a poor job and does not cover up well the data manipulation. One needs to keep in mind that there are literally thousands of fraudulent papers out there. We surely do not know on how many suspicious analyses these forensic techniques have already been tried and so cannot know how good the record of forensic techniques is.

Clever Math

Benford's law, named after Frank Benford (1938), describes a statistical regularity in large datasets: The frequency distribution of the first digit in (some) data follows a straightforward logic with decreasing probabilities as we move from 1 to 9. It is not fully understood why "a large number of very different natural and social processes" (Dieckmann 2007: 321) generate a distribution of data described by Benford's law, a "Benford distribution" (Fewster 2009).

Benford's law suggests that data will converge to this distribution if the process of generating the data satisfies certain conditions and shows no irregularities. Accordingly, if you find data distributions that are inconsistent with Benford's law, you may have found evidence of data manipulation. But this evidence remains tentative: Deviations from Benford distributions can also be caused by the simple fact that the data do not satisfy the conditions that lead to a Benford distribution in the first place. Bummer.

What are the conditions that give rise to Benford's distribution? Most importantly, the application of Benford's law requires a large range in the

Table 10.1
Benford Distribution for the First Three Digits

	1st digit	2nd digit	3rd digit
0	.0000	.1197	.1018
1	.3010	.1139	.1014
2	.1761	.1088	.1010
3	.1249	.1043	.1006
4	.0969	.1003	.1002
5	.0792	.0967	.0998
6	.0669	.0934	.0994
7	.0580	.0904	.0990
8	.0512	.0876	.0986
9	.0458	.0850	.0983

order of magnitude of the observed information. For example, the size of cities fits well with Benford's law, as the population of settlements spans at least eight orders of magnitude. For GDP per capita, which spans four orders of magnitude, it is more questionable whether Benford's law applies.

This requirement does not seem to make Benford's law analyses a promising tool for detecting scientific fraud in regression coefficients and p-values. Variables are often transformed, producing a small range of variable coefficients. The same is clearly true for p-values and significance levels: The range of results is too small to allow a valid comparison of the first digits.

Nevertheless, Benford's law has been used to detect scientific fraud, usually in published data other than reported regression coefficients and p-values. The idea of using Benford's law to detect data fraud can be traced back to Hal Varian (1972). Since then, it has been used with some success to detect fraud in election results (Deckert et al. 2011), campaign spending (Cho and Gaines 2007), accounting data (Durtschi et al. 2004), public official records (Miranda-Zanetti et al. 2019), tax returns (Demir and Javorcik 2020), official COVID-19 mortality data (Silva and Figueiredo Filho 2021), and much more.

Fraud investigators differ in their views on which digit the forensic analysis best focuses on:

1. Analyzing the distribution of the first digit is the most common way to apply Benford's law to any data.
2. Dieckmann (2007) suggests focusing on the second, third, and fourth digits of the coefficients in regression analyses.
3. The Embassy of Good Science suggests focusing on the last digit and expecting a uniform distribution.[3]

Focusing on the first digit has been most common because the first digit has the largest expected deviation from a uniform distribution, and so fewer observations are needed for any deviation to be statistically significant. But what about subsequent digits? Andreas Dieckmann (2007) argues that Benford's law best reveals data fabrication and manipulation in the second, third, and fourth digits. He used the coefficients of published statistical analyses and contrasted their deviation from a Benford distribution. Similarly, he compared and contrasted fabricated regression coefficients that participants in experiments were asked to "invent" with a Benford distribution. Dieckmann finds that for the first digit, both the published coefficients and the fabricated coefficients approximately followed the distribution predicted by Benford's law. However, the second, third, and fourth digits of the fabricated data showed a greater deviation from Benford's law. Dieckmann (2007: 328) concludes that "it is not the first digit that matters! . . . The inspection of the rightmost digits in fabricated data provides better clues to errors or data fabrication than does the inspection of the first digit."

The results of Dieckmann (2007) have to be taken with a grain of salt. Benford's law is not universally applicable and does not necessarily apply to the regression coefficients and standard errors reported. More importantly, even when Benford's law appears to be applicable, any results are only indicative and tentative, not conclusive. Deviations from Benford's law are often random, and the smaller the sample, the greater the uncertainty about the influence of stochastic processes. As a result, evidence that data distributions systematically deviate from Benford's law may raise suspicions, but though tax authorities occasionally detect deviations from Benford's law in tax returns, fraud has yet to be proven by harder evidence. Filing a tax return with too many nines as the first digit does not prove anything.

The main problem with using Benford's law to detect scientific fraud is this: Since the sample size must be large for the analysis to work well, results are more reliable when we analyze a large number of articles than when we analyze the coefficients or p-values of a single article. This is more likely to meet the assumptions underlying Benford's law. Suppose, for the sake of argument, that we analyze the p-values of 1,000 scientific articles and find that the second digit after the decimal point contains 33 percent fours—a tentative indication that p-hacking (or selection on p-values) occurs in about 20–25 percent of published articles. This means that empirical results may have been manipulated in about 200 to 250 of the 333 articles that fall into this category, but nothing tells us which ones. We would need to closely examine all 333 articles to find more conclusive evidence of fraud. Not a promising proposition for systematic fraud detection on a large scale.

Fraud hunters do not do their hard work for glory and recognition and certainly not to get rich. Few become known for the services they provide to the profession. Uri Simonsohn is a notable exception, having played a key role in the downfall of Dirk Smeesters and Lawrence Sanna and now perhaps Francisca Gino. He has also received deserved recognition for his work on p-hacking and preregistration. And he is credited with the most sophisticated, and perhaps most useful, proposal on how to infer tweaking from the distribution of p-values (Simonsohn, Nelson, and Simmons 2014).

However, we start our discussion at the other end of the spectrum. The least sophisticated "detection algorithm" (and arguably the potentially most misleading) is used for testing for the existence of an overrepresentation of p-values just slightly below the significance threshold of 0.05. As Bruns and Ioannidis (2016) suggest, "the intuition is that p-hacked studies manipulate their estimates to achieve p-values that are just statistically significant but really small p-values close to zero are difficult to obtain by p-hacking. Hence, it is argued that true effects can be identified by right-skewed p-curves whereas left-skewed p-curves or a peak of p-values just below 0.05 indicate evidence for p-hacking."

Are p-values slightly below the 0.05 significance level overrepresented? In a large number of meta-analyses across a wide range of fields and disciplines, the answer seems to be "yes" (Head et al. 2015; Assem et al. 2017; Brodeur et al. 2020; Masicampo and Lalande 2012). P-values just below the threshold that separates significant from nonsignificant results are clearly overrepresented in published research. In one meta-analysis, Head et al. (2015) text-mined p-values from more than 100,000 articles across a range of scientific disciplines. They then used a binomial test to compare the number of p-values between 0.000 and 0.025 with the number of p-values between 0.025 and 0.050. Assuming that p-values are uniformly distributed in the absence of p-hacking, Head and her co-authors suggest that p-hacking becomes apparent when more observed p-values fall in the 0.025 to 0.050 category than in the 0.000 to 0.025 category. They conclude that "p-hacking is widespread throughout science" (Head et al. 2015: 1).

The assumption that p-values are uniformly distributed is common among researchers who try to find evidence for p-hacking. At first glance, the assumption that p-values are uniformly distributed if they are randomly drawn from a population appears plausible. At least, this sounds like a plausible assumption to methodologists. They have been trained to believe that a false hypothesis has an a priori 5 percent probability of showing a significant effect. This result crucially depends on the assumption of a uniform distribution of p-values under the null hypothesis (Sackrowitz and Samuel-Cahn 1999).

But are they? Ultimately, there is no simple answer. In a widely cited paper, Simonsohn, Nelson, and Simmons (2014) proposed p-curves as a tool to calculate the probability of p-hacking across a large number of publications, and even in individual manuscripts where the number of p-values may be small. Simonsohn, Nelson, and Simmons (2014) show how the shape of the p-curve depends on the "power" of an analysis, which is a function of the number of observations, the signal-to-noise ratio in the data, and the effect size of the variables (Sackrowitz and Samuel-Cahn 1999). They show that the theoretically expected proportion of p-values below 0.05 can be as low as 2.5 percent, but can also be very large, depending on the size of the estimated effect. As Cohen's d, a popular measure of effect size in psychology, approaches 1, the expected proportion of p-values below 0.05 approaches 100 percent.

The logic is simple. In figure 10.1 we show the theoretically expected p-values from a random draw of 1,000 parameter estimates. On the *x*-axis we have the range of possible p-values, ranging from 0 to 100. The *y*-axis shows the count of the actual number of simulated p-values in each bin, where each bin has a width of 0.025. This gives us two bins with significant

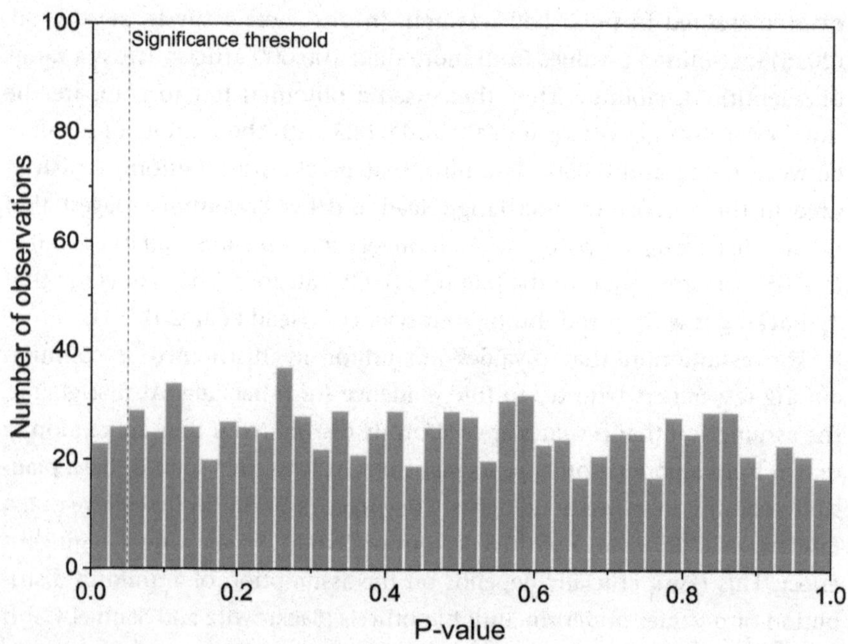

Figure 10.1
Expected stochastic p-value distribution without p-hacking.

p-values to the left of the significance threshold of 0.05. All p-values to the right of this threshold are insignificant. Since we display a random draw from a standard uniform distribution, the number of p-values in each of the two significant bins deviates slightly from the theoretically expected number of 25. In our case, we get 23 p-values between 0.000 and 0.025, 26 p-values between 0.025 and 0.050, 29 p-values between 0.050 and 0.750, and 25 p-values between 0.075 and 0.100.

Let us assume everybody p-hacks and p-hacking exactly halves the p-values. Both assumptions are unrealistic and the second more so than the first. Nevertheless, we need to make assumptions, and with these assumptions we get to a distribution of theoretically expected p-values that we display in figure 10.2, which displays the logic used by Head et al. (2015).

By reducing the p-values between 0.050 and 0.100 by 50 percent, we get more p-values in the second bin ranging from 0.025 to 0.050, while none are left in the third bin that goes from 0.050 to 0.075.

Thus, in a world with theoretically expected distributions of p-values, we can straightforwardly compute the number of p-hacked analysis: Call n_1

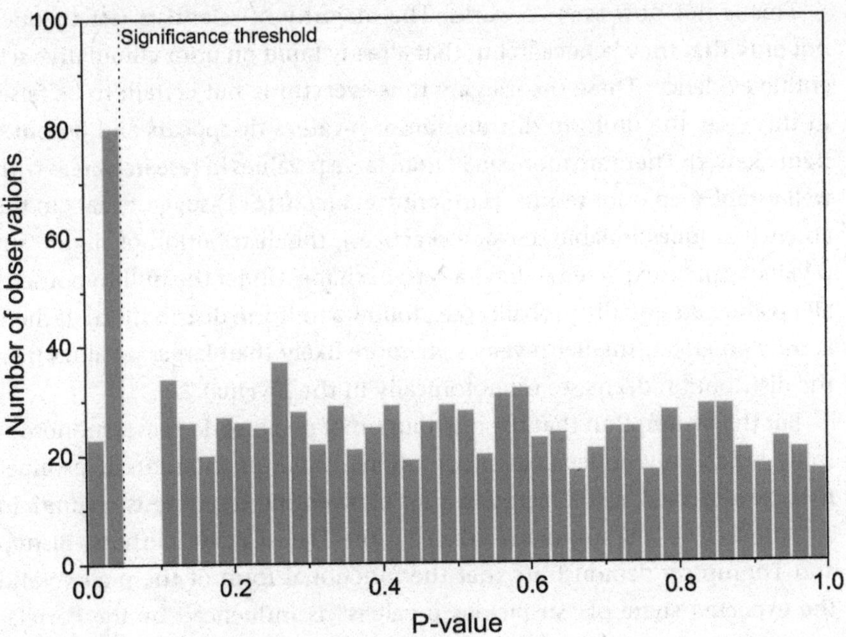

Figure 10.2
Predicted p-value distribution with p-hacking.

the number of papers in the first bin (0.000 . . . 0.025) and n_2 the number of papers in the second bin (0.025..0.050), then we have zero p-hacking if $n_2/n_1 \approx 1$, and as the ratio goes up we have increasingly more p-hacking. Hence, the ratio between the first bin and the second bin p-values allows us to calculate the share of p-hacked estimates. In theory, that is.

But is this wonderful world of theoretical methodology realistic? Are these methodological assumptions relevant to actual empirical research? And does empirical research live up to these methodological assumptions? For Hartgerink et al. (2016: 716), the assumption is intuitively plausible: "Both p-uniform and p-curve are founded on the statistical principle that the distribution of p values conditional on the true effect size is uniform. This same statistical principle underlies standard null-hypothesis significance testing, where the p values are uniformly distributed when the true effect size equals zero." The plausibility of this assumption rests on the strong prior assumption that p-values are uniformly distributed. This means that researchers either have no or use no prior information about the plausibility of statistical associations. No prior knowledge, no theory, no intuition: these are strong assumptions.

This is not how science works. The majority of scientists test theories not only that they believe in but that already build on prior cumulative scientific evidence. These theories are thus everything but certain to be false. In this case, the uniform distribution of p-values disappears and becomes right-skewed: There are more small than large p-values in research areas with well-established prior results. Hartgerink et al. (2016: 1) suggest that "in the absence of [questionable research practices], the distribution of significant p-values can be expected to have a certain shape. Under the null-hypothesis all p-values are equally probable (i.e., follow a uniform distribution). If there is truly an effect, smaller p-values are more likely than larger p-values (i.e., the distribution decreases monotonically in the p-value)."

But the assumption that the distribution of p-values decreases monotonically holds only under certain conditions. What distribution of p-values is to be expected in the actual reality of scientific hypothesis testing? In a comment on the approach taken by the Data Colada authors, Bishop and Thompson demonstrate that the functional form of the p-curve and the expected share of "suspicious p-values" is influenced by the correlation between explanatory variables. In their view, the share of suspicious p-values depends on "(a) the number of studies included in the p-curve analysis; (b) the proportion of studies where ghost p-hacking was used; (c) the number of variables in the study; (d) the sample size and (e) the correlation between variables" (Bishop and Thomson 2016: 9).

The extent of p-hacking cannot be inferred from an overrepresentation of p-values just below 0.05. Furthermore, although p-hacking causes an overrepresentation of p-values just below the critical value of 0.05, the clustering of p-values just below the critical value is neither a necessary nor a sufficient condition for p-hacking. As Bishop and Thompson (2016: 1) conclude: "The absence of a bump in the p-curve is not indicative of lack of p-hacking. Furthermore, while studies with evidential value will usually generate a right-skewed p-curve, we cannot treat a right-skewed p-curve as an indicator of the extent of evidential value, unless we have a model specific to the type of p-values entered into the analysis. We conclude that it is not feasible to use the p-curve to estimate the extent of p-hacking and evidential value unless there is considerable control over the type of data entered into the analysis."[4]

The upshot is this: No assumed and no observed distribution of p-values, and no discrepancy between the two distributions, will ever prove p-hacking—at least not beyond reasonable doubt. What scrutiny of p-value distributions can achieve is identifying a list of articles that together exhibit a suspicious distribution of p-values. But as any police officer will tell you, suspicions can be wrong, and suspicions about a large group of people are bound to be wrong about at least some of the individuals in the group.

Replication, Meta-Analysis, and Robustness Tests

Flaws can be found in any research design. At the very least, any research design, any model can be modified in various ways that are no less plausible than the specification choices made by the original authors. Model uncertainty is inherent in empirical analysis. Two different, but equally plausibly specified models will often produce two different results: results that differ to a greater or lesser extent. One implication of the pervasiveness of model uncertainty is that no single empirical study can be expected to produce results that represent "the truth." As Nosek and his 268 co-authors have suggested, "A single study almost never provides definitive resolution for or against an effect and its explanation. . . . Scientific progress is a cumulative process of uncertainty reduction that can only succeed if science itself remains the greatest sceptic of its explanatory claims" (Open Science Collaboration 2015: 4716–4717).

If a single study cannot be trusted because equally plausible alternative analyses can produce different results and support different conclusions, will multiple analyses of the same, or at least very similar, hypotheses, which produce the same, or at least very similar, results, do the trick?

Science has always relied on multiple analyses of the same hypothesis. If a hypothesis, a theory, a prediction is interesting enough for one researcher to work on, there are likely to be other researchers following suit. A growing field of research will almost inevitably become diverse in research designs and model specifications, and the growing variation in model specifications and research designs often leads to different, occasionally inconsistent empirical results. But what does a large variance in results and inferences tell us about individual studies and the likelihood that they are based on manipulated or fabricated data or tweaked analyses? Three relevant techniques have been used to provide tentative evidence: replication analysis, meta-analysis, and robustness testing.

Exact replication analyses rerun the same model on the same data. Exact replication will produce identical results unless the analyses are flawed or a random component is repeatedly drawn into the analyses. But even if this happens, we would still expect statistically indistinguishable results for the majority of repetitions. In the absence of random components, and if the replication analysis holds the dataset and the command lines (do-file) constant, replication results have to be identical to the original results.

As an editorial in *Nature Neuroscience* suggests, "although replication is an essential part of the scientific enterprise, in practice few experiments are replicated exactly. This problem is compounded in the biological sciences because biological variability can explain away discrepant results. Replication of results can help validate the science when it is suspect, but failure to replicate can rarely be used as proof of fraud; it is common for two honest scientists to get different results from the same experiment" (*Nature Neuroscience* 2006). In fact, nothing limits this observation to "biological variability." Any probabilistic causal relationship generates variability in empirical results, which is expected and unavoidable. Replication analyses can only provide tentative evidence for tweaking, fabrication, or manipulation because failure to replicate (failure to obtain the same or very similar results) can have any number of causes.

Both replication analysis and robustness tests are more informative if either the sample or the model specification or research design does not exactly match that of the original analysis. The famous replication analyses of over 100 studies in psychology by Open Science Collaboration (2015) did not exactly replicate the findings but reanalyzed these studies with new experimental data or updated observational data. Because of these changes, results will inevitably vary from those of the original articles. Factors that influence the (dis-)similarity of original and replication analyses include, but are not limited to, sample, setting, statistical power, and procedural variations (Klein et al. 2014: 143).

However, the results of replication or robustness tests will inevitably deviate unsystematically from old results unless either the old or the new, updated results have been carefully selected. In Nosek et al.'s 100 replications of highly cited studies in social psychology we see systematic deviations of the replication results from the original results—systematic deviations that indicate that the original results have been selected either by their authors or in the review process.

Meta-analyses are designed to look at the variation in the results of the same or very similar hypothesis tested in a large number of studies. Meta-analyses can identify which studies produce results that stand out from the mainstream of studies. Meta-analyses have been criticized for comparing apples to oranges. They often combine estimates that only appear to be studying the same effect but often are incomparable because the treatment itself or the outcome are not sufficiently similar.[5] If this is the case, nothing can be learned from apparently outlying results because they should not be included in the same meta-analysis in the first place.

Similar to replication analyses, a study may stand out from other studies for various reasons. Tweaking, fabrication, or manipulation are among these causes, but without further analysis it is not clear why the results of a seemingly outlying study differ from those of other studies. Meta-analyses reveal outliers, and while these outliers can be the consequence of a deliberate manipulation of empirical results, they can also fall within the natural variability of probabilistic causal relationships. There is no a priori reason why outlying results in a meta-analysis must be inferior to other studies that are not outliers. Empirical analyses are not independent of each other; elements of research design and key assumptions made in a model specification usually carry over from early to later analyses. Nothing guarantees that the assumptions made in earlier analyses are correct. As a result, the majority of analyses in a meta-analysis may well be biased, while the minority of studies that might be considered outliers are actually closer to "the truth." Even if this does not hold, meta-analyses will not detect fraudsters who ensure that their results remain within the mainstream. Many fraudsters make a point of publishing results that stand out from the crowd in order to attract attention. But others do not, and these cannot be detected by meta-analyses.

Robustness tests re-estimate the original analyses but with plausible changes to various aspects of the model specification. The idea is that results from the baseline model have potentially been tweaked if they are fragile, sensitive, or not robust to plausible changes in model specification. Exactly what these terms mean is poorly defined and can range from a change in the sign (direction) of an effect size to a change in the statistical significance of an effect size to a statistically significant difference in effect sizes. As we explain in detail in our book on robustness testing in empirical research, we prefer to quantify robustness by the degree of overlap of the confidence intervals between the baseline model and a robustness test model, a continuous measure that goes from 0 to 1 (Neumayer and Plümper 2017).

A lack of robustness to plausible changes of the model specification tentatively indicates that the authors have selectively chosen a model specification that has provided them with their results. Lack of robustness tentatively indicates tweaking. In the past, researchers could reasonably claim that they were unaware of the lack of robustness of their chosen (baseline) model specification. This defense has become hard to maintain because a robustness testing revolution has swept the social sciences over the last two decades or so, which has led authors themselves to provide the results of a large number of robustness tests to accompany their baseline model results, at least in traditional regression analyses of observational data. It is true that authors are likely to report only robustness tests that are favorable to their baseline model and to suppress the presentation of robustness tests that produce deviating results. As a consequence, robustness tests that have been designed by others may still find substantively different results, even though the authors of the original article suggest that their results are robust. Thus, extensive sensitivity to model specification assumptions provides only tentative evidence for fraud.

Text Mining: Do Fraudsters Write Differently?

Do fraudulent authors reveal themselves because they write differently?

At first glance, this idea seems implausible. Intuitively, there seems to be no reason to believe that fraudsters write differently. Sometimes, however, intuition can lead one astray. Fraudsters are aware that they are cheating, and psychological dynamics can influence how they write up their fraud, even if they are not aware of these dynamics. In a paper published in *PLOS One*, Markowitz and Hancock (2014) took the single-authored articles published by Diederik Stapel, a notorious serial fraudster with whom readers will be quite familiar by now, and divided them into two categories: retracted and not retracted. They then calculated how often these two categories of articles used different discourse categories. They found that Stapel used terms related to methods, investigation, and certainty (such as "explicit," "certain," "definite") much more often in his fraudulent articles than in his "genuine" articles (not retracted articles were assumed to be genuine, and retracted articles were assumed to be fraudulent). There were statistically significant differences in the use of other categories, but with 360,000 words analyzed, some differences become statistically significant, though they remain substantively irrelevant. For example, Stapel used words like "feedback," "assess," and "research" about 15 percent more often in his fraudulent articles, and this difference is significant, but this 15 percent difference does not help us to identify fraudulent articles with sufficient certainty.

To Markowitz and Hancock (2014: 2) the case is clear-cut: "Stapel's fraudulent writing featured significantly higher rates of terms related to scientific methods and empirical investigation compared to his genuine writing, while cause and effect terminology and quantities did not differ across the two corpora. These data suggest that fraudulent papers involved the overproduction of scientific discourse, such as terms related to explaining data and research processes." They even claim that their research design managed to hold other factors constant: "The patterns are impressive given that the only difference between the two corpora was the fact that they reported on fraudulent data. In many other respects they were identical, including each being written by the same first author and each focusing on topics within social psychology" (Markowitz and Hancock 2014: 3).

Is the technique developed by Markowitz and Hancock (2014) promising for detecting fraudsters? The authors correctly categorized 71 percent of Stapel's papers as either fraudulent or honest—assuming, of course, that retractions are a perfect indicator for separating Stapel's published articles into fraudulent and honest. At a first glance, this seems to be a success, but the results are actually closer to random categorization than to perfect separation and not good enough as a technique for detecting fraudulent papers. The apparent success rate of 71 percent correct categorizations could be the result of confounding factors or model misspecifications. Stapel did not start out as a fraudster but became one over time. This means that, on average, his fraudulent articles were written and published later than his honest work. Stapel may have changed his writing style over time or the content of his articles may have changed.

Our main concern, however, is a different one. Even if Stapel used certain scientific-sounding terminology disproportionately in his fraudulent works, it does not follow that articles that use a disproportionate amount of scientific terminology are fraudulent, nor does it follow that other fraudulent researchers will adopt the same pattern of behavior. Induction is not deductively valid.

Interestingly, however, Markowitz and Hancock (2016) follow up on their earlier paper, this time looking at 253 papers retracted for data fraud from a variety of authors and disciplines, and comparing them with 253 matched unretracted papers, as well as 62 papers retracted for reasons other than fraud. They find that fraudulent papers "were written with significantly higher rates of linguistic obfuscation, including lower readability and higher rates of jargon, than unretracted and nonfraudulent papers" (Markowitz and Hancock 2016: 435).

Pixel Peeping: Duplicated and Manipulated Images

The short list of famous fraud hunters certainly includes Elizabeth Bik, who is almost famous for her attempts to put a very specific class of fraud on the agenda. Bik was, to our knowledge, the first scientist to bring the issue of fraud in scientific images to the limelight.

She began her new career as a "pixel peeper" by scrolling through a dissertation that she thought might have been plagiarized. It turned out that she found a different kind of self-plagiarism: Two papers by Debasmita Mandal and Alan Levine (one of these articles is co-authored by Pingfu Fu) used a single image of proteins, suggesting that images came from different experiments.[6]

Bik discovered that the upper right image shown in figure 10.3 is a moderately stretched version of the left image, while the lower right image is inverted and tilted.[7] Of course, the duplication of an image does not prove scientific misconduct. But stretching, tilting, and flipping a single image suggests that the authors deliberately tried to cover up their multiple usages of a single image. And that is a red flag.

Bik became curious. Is pixel-hacking widespread in scientific publications? She selected a sample of articles published in *PLOS One*. *The New Yorker* reports that she quickly went through circa 100 articles and found "a few duplicate images."[8] Honest mistakes or scientific misconduct? Many of these images were transformed—stretched, mirrored, flipped, tilted—before they were used more than once in a single publication. And there were too many of them to be explained by simple mistake and careless oversight. In an opinion piece for *The New York Times*, Bik suggests that science has "a nasty photoshopping problem."[9]

Today, Elizabeth Bik describes herself as a scientific integrity consultant and runs the *Science Integrity Digest*. According to this website, as of November 2023, Bik's work has resulted in "1069 Retractions, 149 Expressions of Concern, and 1008 Corrections," and counting.

In late 2022, Bik became involved in an evaluation of Stanford's University President Marc Tessier-Lavigne. Tessier-Lavigne was accused of being the co-author of articles that contain "multiple altered images." The investigation was triggered by Theo Baker, a Stanford student and an author who turned a student-run journal, the *Stanford Daily*, into a respected source that

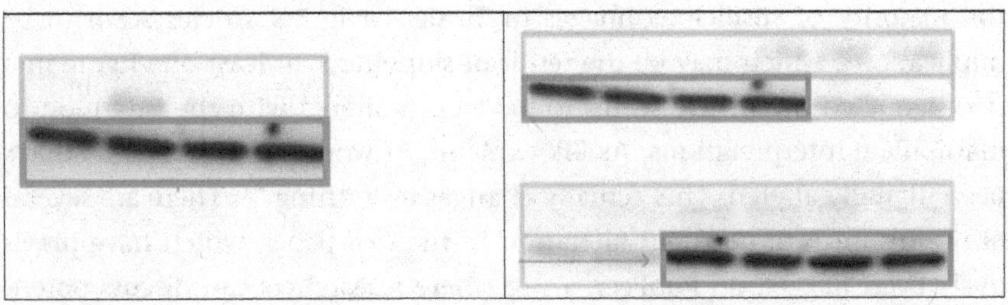

Figure 10.3
Image self-plagiarism detected by Bik.

won the prestigious George Polk Award. Baker himself also won a "special award" and was invited to write an article for *The New York Times*. Occasionally, it pays to blow the whistle. As the story goes, Baker asked Bik to take a closer look at some images that he thought were duplicates and manipulated. According to Bik, these and other articles published in *Science* and *Nature* contain "serious problems."

Stanford University investigated the case and eventually published a report that led to the resignation of its own president. Wow. Nevertheless, parts of the report appear defensive and to be making the problems with Tessier-Lavigne's work seem smaller than they actually are. On the one hand, Stanford seems to hold him accountable for what went on in his lab: The report suggests that Tessier-Lavigne "created a laboratory culture with many positive attributes, but the unusual frequency of manipulation of research data and/or substandard scientific practices from different people, at different times, and in labs at different institutions, suggests that there may have been opportunities to improve laboratory oversight and management."[10] But then the investigation concluded that "for the five reviewed papers where Dr. Tessier-Lavigne was a principal author. . . . Dr. Tessier-Lavigne did not have actual knowledge of the manipulation of research data that occurred in his lab and was not reckless in failing to identify such manipulation prior to publication." Apparently, a lab director is to be held accountable for the lab culture, but is not reckless in failing to identify the falsifications and manipulations that this lab brings about.

Tessier-Lavigne has a track record of sloppy research practices. In 2015, he had to work on corrections of some articles published in *Science*, but apparently the journal did not publish the corrections. In 2023, he decided to retract the faulty articles. But "sloppiness" is not the whole story. While the majority of suspicious images in Tessier-Lavigne's articles seem to be duplications, which may be the result of sloppiness, at least one figure in a now retracted *Cell* article seems to have been altered with the intention to manipulate interpretations. As Bik explains, "I would testify in court that's been digitally altered. This actually changes everything."[11] There are several other instances of apparent alteration in the *Cell* paper, which have previously been flagged on PubPeer, a site where researchers can discuss potential issues in papers. Nevertheless, the university told the *Stanford Daily* that smaller errors initially described by the student paper in its first piece had been brought to Tessier-Lavigne's attention in 2015, and that *Cell* decided a correction was not necessary.

Tessier-Lavigne's reaction to these irregularities came too late and his defense was too weak. The *Stanford Daily* continued to publish expert interviews on manipulated images in the work of Tessier-Lavigne, and the fraud scandal attracted wider attention perhaps motivated by the fact that Tessier-Lavigne was Stanford University's president. Eventually, he resigned effective August 31, 2023.[12]

As interesting as the Stanford case and the role of image duplication and manipulation are, how common is the problem and how promising is scrutinizing images as a tool for detecting fraud?[13] In a first attempt to quantify the severity of the problem, Bik, Casadevall, and Fang (2016) analyzed more than 20,000 biomedical papers and found that around 4 percent of articles used duplicated images inappropriately. Given that images can be manipulated in more complex ways than duplication, and therefore would not necessarily have been detected by Bik et al. (2016), this is alarming. Bik suggests pre-publication image screening, undertaken by professionals and paid for by academic journal publishing houses, as the way forward.

Duplication is the lazy approach to image fraud, just as data fabrication is the lazy approach to getting the quantities of interest you want. Duplication is the easiest form of image manipulation to detect. Less lazy fraudsters can manipulate images in ways that cannot be detected. In the world of Adobe's Photoshop, getting rid of the telltale dot above the image in figure 10.3 is easy. Artificial intelligence systems such as DALL-E will soon be able to produce scientific images that are difficult, perhaps impossible, to detect.

Figure 10.4 shows what can be achieved in less than five minutes using a graphics editor. The spot that gave away the duplication has been removed and the contrast has been improved. In another five minutes, one can change the shapes in the image, the perspective of the image, and flip the

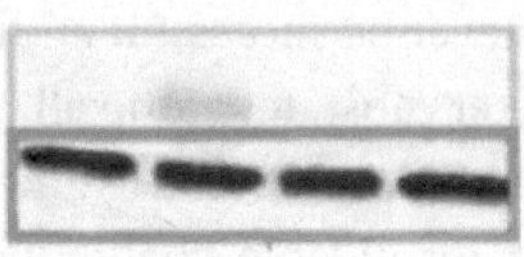

Figure 10.4
The image after five minutes of manipulation with a graphics editor.

image. The resulting image would still have similarities to the original, but the new image would hide the duplication much better and perhaps to the point where only more advanced forensic pixel-peeping strategies can detect the duplication.

Scientists modify images all the time, as scientific journals, reviewers, and editors pay a premium for pretty images that support a paper's argument in a striking way. Such "beautification"[14] of scientific imagery is different from image manipulation, which attempts to bias the interpretation of results in the direction desired by the researcher, or to provide evidence for an argument that is not actually supported by empirical evidence. NASA's image gallery provides a good example of image beautification. These stunning images of the universe are, well, manipulated. NASA routinely photoshops raw images "to enhance details or to visualize things the human eye could never see." This is not scientific fraud unless these "enhanced" images are used in scientific publications and the enhancements deliberately manipulate the way readers interpret these images. However, the boundaries between beautification and manipulation are blurred and imprecise.

Pixel peeping is an uphill battle that is unlikely to be won given that technologies exist that can alter every single pixel in a digital image. Artificial intelligence may still be at an early stage of development, but technological progress has already reached the state that enables AI to produce digital images that are difficult to distinguish from real images. In the not so far future, it will no longer be difficult but impossible to tell one from the other. Fraud hunters who search for manipulated pixels have their future work carved out.

Does Detection Work?

To deter scientific fraud, fraudsters need to believe that their unethical behavior will be detected with a high probability. We suspect that current detection techniques will never be successful enough to increase the risk of detection sufficiently to deter most, let alone all, potential fraudsters.

Successful detection must simultaneously meet four conditions, each one of which represents a formidable challenge on their own and even more so in combination:

1. Fraud detection strategies need to identify *individual* articles that contain manipulated or heavily tweaked empirical evidence. It is not sufficient

to estimate the prevalence of the problem in a field or across a series of studies.

2. The number of false positives needs to be very close to zero. Science suffers a serious blow if articles that later turn out to be innocent are retracted for data fraud.
3. The number of false negatives must be small enough to ensure that fraudsters have a sufficiently high probability of being detected. If a few fraudsters get away with manipulated results, so be it, but the number needs to be much smaller than it is now.
4. Detection must be strategy-proof. This means that potential fraudsters must not be able to continually find a way around new detection techniques.

Existing techniques for detecting manipulated empirical results work occasionally. One can always find anecdotal evidence that these techniques have successfully detected some fraud somewhere some of the time. But we only know about successful detections; by definition, we never hear about the undetected cases. The size of the invisible part of the "iceberg" of undetected fraudulent studies will never be known. The race between fraudsters and those who seek to prevent or at least detect fraud cannot be won by the latter. But giving up is not an option. Existing detection techniques cannot reliably detect all manipulated empirical results. So far, the vast majority of research into fraud detection has been carried out by self-motivated scientists who felt the need to do something about the problem. Few have become famous. An exception that proves the rule is Brian Nosek, the corresponding author of the Open Science Collaboration replication exercise. But who remembers the more than 100 other researchers who contributed to the collaborative replication effort? People like Peter Attridge, Grace Binion, and Sara Steegen, to name just three of Nosek's collaborators? Likewise, Uri Simonsohn is deservedly famous, but what about the others?

Detecting and fighting fraud is a lonely and under-rewarded business, and there is a clear need to incentivize and institutionalize the effort. No white knight will come along, however. Funding agencies may have representatives and chairs who, in Sunday speeches, stress the need to do more to defend the scientific project against the fraudsters who put their academic careers ahead of the principles of scientific rectitude. No institution monitors the integrity of scientific publications. No funding agency provides the resources to make the systematic application of forensic statistics, replication, meta-analysis, robustness testing, or image screening worthwhile. Even funding for data repositories seems to come from universities rather than funding agencies. Highly profitable publishing houses do not enforce data transparency rules for their journals and do not pay for exact replicability checks or checks for image duplication and manipulation.

All this suggests that a focus on fraudulent intent is not promising. That is bad news. But it also may not be necessary, and this is the good news. Bik does not infer intent when she looks for and finds manipulation and duplication. Nor do the majority of fraud hunters. An analysis of fraud chasers finds that they "[do] not seek to assess or prove intent when assessing the reliability of study results" (Parker et al. 2022: 11). Whether results are trustworthy or not matters most, and not whether errors, omissions, or biases were intentional or the mere result of honest mistakes. Fraud hunters primarily seek not to expose fraudsters but to identify errors, irregularities, deviations from good scientific practice.

Communicating the uncertainties inherent to empirical results, not overstating the results of a paper, not politicizing these results, is key to preserving and restoring trust in science.

11 Two Steps Forward

When in doubt, know your way out.
—Zombieland Rule #22

The Nature of the Problem

Science is not broken. Deterrence, prevention, and detection strategies do work, at least to some extent. Every once in a while, a fraudster is caught, and if the evidence is sufficient, the fraudster is banned from the community of honest scientists. Justice is served, but the problem persists. It seems that science must live with the inconvenient truth that journals publish papers with manipulated empirical results, that scientists who have manipulated empirical results not only get away with it but receive promotions, research grants, and their seven minutes of fame.

It is true that there will always be some scientific misconduct, unless the scientific community were willing to stop publishing research results before independent replication studies and meaningful robustness tests have by and large confirmed the original results. This is a remedy, but one that would unduly slow down the research process and therefore be potentially more costly than the damage caused by fraud and tweaks.

Still, scientists should not shrug their shoulders and assume that there is little they can do beyond what is already done. The deliberate manipulation of empirical results is a serious problem: perhaps rare, but certainly serious. The "do nothing" strategy is tempting. It would even seem to be a sustainable approach, unless there were a significant shift in the willingness of governments to fund agencies such as the National Science Foundation, the European Research Council, the UK Research and Innovation, or the Deutsche Forschungsgemeinschaft. Given that the modern world, with its high levels of material well-being and human development for billions of people, is inconceivable without science, the credibility crisis does not really threaten the continued existence of science in its current form. But the credibility crisis of science is corrosive, undermining the scientific project and chipping away at trust in science one fraudulent paper at a time. Therefore, it is the responsibility of scientists "to figure out how to build a more trustworthy, credible scientific base."[1]

This book has argued that variants of scientific fraud exist that cannot be proven beyond a reasonable doubt. In this chapter, we suggest strategies that, against all odds, aim to reduce the manipulation of empirical research results. These strategies have limitations. An ultimate remedy to the credibility crisis of science is likely to have costs that exceed the costs of the crisis itself and ultimately damage the production of scientific knowledge and ideas.

Scientific fraud differs from murder in a way that we have not yet mentioned: In the vast majority of cases, murder leaves a dead body on the floor and the detective only has to provide an answer to the "who did it" question. With scientific fraud, this question can be answered easily, and the answer is: the authors, who are responsible even if they have left the pushing of the button to a lab assistant or a research student. The question that fraud chasers need to answer is whether data have been manipulated or fabricated. Those who try to prove scientific fraud resemble the detective who finds a smoking gun and now has to prove that somebody has been killed. But not all smoking guns have killed somebody. Likewise, when empirical results appear too good to be true, when a researcher published too much in too little time in too good journals, they might just be a superstar in the making. "There is no smoke without fire" does not hold for scientific fraud, and, at the same time, there are many fires without smoke.

And there is an elephant in the room. Tweaks are serious but, as far as we know, not rare. As Uri Simonsohn told Christie Aschwangen, "everybody has p-hacked at least a little bit."[2] We have no idea how he could possibly know, but we definitely cannot prove otherwise, and he is probably right. Evidence of tweaking is hard to bring by; the absence of tweaking is impossible to prove. To make matters worse, science has yet to find means to effectively tackle McIntyre's "murky category." Tweaking is not harmless. It has identical or at least very similar effects on the fabric of science, on trust in science, on the competition between scientists as fraud, but where data fabrication and manipulation are an open onslaught on scientific integrity, tweaks are an ambush in disguise.

Downplaying the damaging impact of tweaking on the credibility crisis of science does not help solve the crisis. Science needs to focus on the deliberate manipulation of empirical results, regardless of the skill with which fraudsters get the results they want or need: The smarter fraudsters resort to tweaking; outright data fabrication and manipulation, which are quick and dirty, are for the unskilled, the brazen, or the insatiable. Tweaks are fraudulent and have a devastating effect on scientific integrity. Scientists not only cooperate and collaborate; they also compete for scarce resources. When some scientists manipulate empirical results, it increases the pressure on other scientists to do the same.

The fight against the manipulation of empirical results is largely complicated by the dilemma that tweaking is often indistinguishable from genuine model specification decisions and specification searches, but at the same time specification searches are a necessary element of empirical analyses. Tweaks and specification searches merely differ in respect to the (unobservable) intention of the researchers: Do they want to get their model specifications right or are they primarily interested in getting their manuscript past reviewers?

The Burden of Proof

The law is simple: Suspects are innocent until proven guilty beyond a reasonable doubt. Suspects do not have to prove their innocence; the prosecutor has to prove their guilt. Legal practice is often more complicated, and "beyond reasonable doubt" is a term open to interpretation, but that is another story. Or maybe not . . .

Scientists must be able to prove that their data are not fake, provide code that shows how the raw data have been altered and transformed, and provide code that allows the exact replication of the empirical results. These rules of good scientific practice do not reverse the burden of proof, but they do specify the meaning of *beyond reasonable doubt*.

When Michael LaCour could not provide the raw data for the article published in *Science*, the presumption of innocence was no longer reasonable because he could not prove that the survey had ever been conducted. And Harvard University's 1,288-page report on scientific misconduct by

Francesca Gino provides ample evidence of data manipulation beyond a reasonable doubt. Under these circumstances, Gino must prove that she did not manipulate the data, which is a seemingly impossible task given the compelling evidence we have seen.

It is important to understand that scientists are allowed to alter and transform raw data. In fact, these manipulations are part of the research process. However, unlike what fraudsters often do, these manipulations must be transparent and justifiable. Scientists must be able to show how they manipulated the raw data, explain why they manipulated the raw data in that particular way, and provide a replication file and the replication data that allow the exact estimates presented in the article under review to be replicated. It is also good practice to demonstrate the extent to which these manipulations affect the empirical results. To make data transformations transparent, an unaltered copy of the raw data must be available to researchers who wish to replicate the data.

Data Transparency and Integrity

Data fabrication and manipulation rely on privately collected data. While it is certainly possible to manipulate publicly available data, the risk of detection is high when the data are public and negligibly small when the data are private. Scientists can only prove the integrity of the data they use if they are able to show that they have not fabricated the raw data on which their analyses rely and able to transparently lay out exactly what alterations and transformations they have undertaken to move from the raw data to the data used in estimation.

Proof that the raw data are not fabricated is easiest to bring by if the data are collected by a public agency or an independent survey company or an independent lab, if these agencies and companies keep and safely store a record of the raw data. Proving that data have not been fabricated if authors collect their own data is harder, but not impossible. In these cases, data collection needs to be outsourced to someone else, for example, to other researchers, or if that is not feasible, the data collection process should be filmed and the footage made publicly available along with the replication data. Of course, we understand that filming is an intervention that may affect participants and their behavior, but it can be done in an unobtrusive way. Some universities have already established a mechanism that allows researchers to upload their raw data without being able to alter it thereafter. Ideally, however, raw data are safely stored in publicly accessible data repositories.

What if authors are unable to prove the integrity of their data?

Scientists who are either unwilling or unable to do so should inform the editor of the journal to which they submit a manuscript. We believe that, if authors are unwilling to provide access to their data for replication purposes, their manuscript should not be published. But of course, such a policy needs to be implemented by the publishing houses and the journal editors.

To improve the transparency on how the data were collected and the replicability of the analyses we propose a data transparency classification scheme (see table 11.1). We do not consider a data transparency classification scheme as a substitute for a reproducible workflow (Christensen et al. 2019). Our idea is that journals should clearly explain the data transparency for all articles they publish. Whether or not they rely on our index is beyond our influence.

Our proposal includes that journals should clearly state the data transparency index scores of the articles they publish. Journals may decide to only accept manuscripts with data transparency index scores above a certain grade, either in the aggregate or on specific scores. This is up to them.

This policy allows articles to rely on data whose integrity cannot be proven, but still be published. We hope and expect that peers will read these articles with healthy skepticism given the lack of data transparency. We see important benefits to readers having full information about the data transparency standards of the articles they are reading. Readers should have more confidence in the empirical results if the data used are consistently of the highest quality than if they are of a much lower quality.

Replicability and Robustness

Replicability is an essential feature of scientific research: the cornerstone of science (Kelly 2019: 1). If empirical results cannot be replicated, the conclusions are not informative and cannot be trusted. When an empirical result can be exactly replicated using the original data and a do-file or similar execution code, this is also referred to as reproducibility. Many journals now check for reproducibility during the review process and do not publish manuscripts whose results cannot be reproduced. These journals require authors to provide the replication data and the code that generated the reported results. Conducting these rigorous pre-publication reproducibility

Table 11.1
Data Transparency and Integrity Scale

A1: The raw data used in this article have been collected by	
Public agencies	a
An accredited company	b
Other researchers	c
The authors of the manuscript	d
A2: The raw data can be obtained from	
Official webpages	a
Institutions, organizations, agencies	b
Other scientists	c
The authors of this manuscript	d
The raw data cannot be accessed	e
A3: The raw data in the article can	
Be manipulated by the author	a
Not be manipulated	b
A4: Authors scientists are allowed to use the raw data used in this article for	
All purposes	a
Replication and reanalysis	b
Exact replication	c
None of the above	d
A5: Computer code that transforms the raw data into the data used in the reported analyses (replication data) can be accessed at	
Data repository	a
Private web page	b
Neither	c
A6: The final data used in the analyses (replication data) reported in this manuscript can be accessed at	
Data repository	a
Private web page	b
Neither	c
A7: The final data used in the analyses can be used for	
All purposes	a
Replication and reanalysis	b
Exact replication	c
None of the above	d

tests is expensive, but it prevents journals from publishing spurious analyses. A smaller number of journals have also begun to check images for questionable practices such as inappropriate image duplication. Again, this is costly, but academic publishers are among the most profitable businesses in the economy.[3] In contrast, replicability explores whether very similar results can be generated based on the same research design and model specification but with new and different data. We should not expect the results to be identical, but the difference should be within the sampling distribution of the original results. Robustness checks are something entirely different. They test the sensitivity of the original results to plausible changes in the research design or, more commonly, in the model specification. Narrow robustness holds the sample constant; broad robustness varies not only the model but also the sample.

Table 11.2 distinguishes between two changes in the analysis: a change in the data (a different sample) and changes in the model specification—a term we use to encompass different assumptions made in the analysis about the process generating the data, including the set of control variables, the dynamic specification, unit and causal heterogeneity, and other assumptions. Robustness is thus more informative than replicability. Reproducibility and replicability accept the research design and model specification of the original analyses, while robustness testing assumes that the research design and model specification are uncertain. To analyze whether this uncertainty potentially invalidate inferences, the results of the baseline model should be tested against alternative plausible designs and specifications (Neumayer and Plümper 2017). Additional insights can be gained by using a sample different from that of the original study. This test is more important for research designs that select samples.

It is time to comprehensively check for robustness because reproducibility and image screening checks do not pose insurmountable hurdles

Table 11.2
Reproducibility, Replicability, and Robustness

		Data	
		Original sample	Different sample
Model specification	Identical	Reproducibility	Replicability
	Different	Narrow robustness	Broad robustness

for skilled fraudsters. Robustness tests can be a powerful tool for assessing the true uncertainty associated with empirical analyses. However, this purpose is defeated when robustness tests are selected by the authors of a manuscript, which is the rule rather than the exception. Authors typically select those robustness tests that confirm the results of their principal or baseline model. Crucially, therefore, the necessary next step in academic publishing is that editors and reviewers exercise control over the selection of robustness tests published with the original article: "We do not argue that cherry picking . . . does not exist with robustness tests. But the severity of the problem should be reduced if reviewers and editors increasingly demand additional robustness tests from the authors or if it becomes more common . . . to replicate . . . the findings of important publications" (Neumayer and Plümper 2017: 230).

Robustness tests are a powerful tool for assessing the true uncertainty associated with empirical analyses. However, this purpose is defeated when robustness tests are selected by the authors of a manuscript, which is the rule rather than the exception. Currently, nothing prevents authors from reporting only robustness tests that confirm the results of their baseline model. To overcome or at least mitigate this problem, we suggest that editors and reviewers should be proactive and take interest in and influence the selection of robustness tests. If editors and reviewers (rather than authors) select robustness tests, the reported results are likely to be more informative about the true sensitivity of the empirical results to plausible changes in the model specification. This is desirable. It is better to acknowledge the true uncertainty around empirical results than to pretend that "the truth" has been found and that all results are robust. Revealing and discussing uncertainty can help identify important future research agendas and open up fruitful new avenues of research. Accordingly, science is likely to advance more rapidly when interesting research results are published, even when the results remain more uncertain than usual.

Elsewhere (Neumayer and Plümper 2017) we have provided numerous examples for useful robustness tests. The optimal set of tests, however, depends on the analyses reported in each individual manuscript. Indeed, the optimal robustness tests may vary significantly from manuscript to manuscript. We have frequently been asked to identify a small set of standard robustness tests that are always useful. We do not think that providing one set of standard tests is feasible. However, there are a few rules that one can use to identify informative robustness tests:

1. In Neumayer and Plümper (2017) we distinguished between model variation, randomized permutation, structured permutation, robustness limit, and placebo tests. Robustness limit tests (or Rosenbaum bounds tests, as they are also called), which directly try to explore where the limits of robustness are if one changes a particular model specification in a particular direction, are most informative. These tests do not ask whether or not results are robust to alternative plausible specifications but ask how much model specifications need to be changed to give results that differ in a predefined way.
2. We recommend always examining the sensitivity of the estimates to dropping influential cases, cases that have a large influence on the effect, from the sample. In the Holden (2016) example that we have discussed in chapter 7, we needed to drop 12 schools from the estimation to render the predicted effect of school book subsidies insignificant: 12 from 2,350 schools in the original dataset. This is useful information because it suggests that Holden almost certainly overgeneralizes his findings. Along similar lines, we only need to drop 2 hurricanes from the sample of 92 hurricanes to make the predicted effect of hurricane names on hurricane mortality insignificant. And we need to reduce the participants in the dictator game from 10,320 to 6,440 to find that men are more altruistic than women. What this tells us is that people are heterogeneous.[4]
3. The uncertainty of model specifications results from simplifying assumptions about the tested theories or about the data-generating process. For example, the majority of social and life science research makes the simplifying assumption that causal effects are identical for all units of analysis. We recommend exploring the robustness of results toward relaxing this assumption, and in Neumayer and Plümper (2017) we show how this can be done.
4. Where relevant, studies should explore the sensitivity of results to plausible changes in the set of control variables and their operationalization.

Robustness tests lessen the problem that tweaking cannot be proven beyond reasonable doubt. Yes, tweaking typically cannot be proven, but scientists can nevertheless explore the *tweakability* of data. The smaller the robustness of empirical results, the larger the uncertainty associated with specification choices and the easier the model can be tweaked to give the desired result. Tweakability essentially represents how easy it is to arrive at a different result. Robustness tests have been invented for a different purpose: they explore the fragility of one's chosen baseline model to plausible changes in specification. However, they can also be used to demonstrate the fragility of results to these changes. And high fragility equals high tweakability.

If an empirical model is not robust, it is tweakable. The extent to which results can be tweaked varies considerably: Some results can be turned on their head by a seemingly minor tweak. Turning an insignificant into a significant effect is often not a difficult task. Some model specification changes can even change the sign of a hitherto statistically significant result in the opposite direction, from positive to negative or negative to positive. In all cases the crucial question is whether the model specification changes that bring about these changes are plausible. No model is robust to highly implausible changes in its specification.

Still, a low level of robustness, while an important signal in itself, does not provide sufficient evidence for tweaking. Tweakability differs from actual tweaking. Data and models can be highly tweakable, but that does not mean that the authors have tweaked. Sensitive empirical results may well be an unavoidable consequence in a particular area of research. A low number of cases and a poor signal-to-noise ratio in the data inevitably lead to high fragility. Lack of robustness mainly indicates the relatively low level of confidence science can have in results. Lack of robustness does not indicate that authors have cheated and tweaked their results.

High tweakability, however, can seduce those with the intention to tweak to actually do so, just as owning a fast car seduces those with a tendency to drive fast to exceed the speed limit or, if on the Autobahn, to go as fast as their car can drive. And just as those with a tendency to drive fast will buy fast cars, those with a tendency to engage in tweaking may choose research areas and projects that offer a high degree of tweakability. In fact, relatively unskilled tweakers will often be attracted by a high degree of tweakability. Skilled tweakers, on the other hand, do not need high tweakability. They need only some tweakability. As a consequence, skilled tweakers may prefer research areas with relatively little, but sufficient tweakability. Only unskilled tweakers need high tweakability to come to the desired result.

A Disclaimer

These proposed measures taken together address the credibility crisis of science by making the stability and robustness of empirical findings the central question and by allowing readers to quickly access the quality of the data used in the article. Christie Aschwanden puts it like this: "To make science more reliable, we need to adjust our expectations of it. . . . Science is not a magic wand that turns everything it touches to truth. Instead, science operates as a procedure of uncertainty reduction."[5] We do not always agree with Christie Aschwanden but here we do.

Our proposed measures are useful, but, and here is the disclaimer, they fail to solve fishing and harking strategies. Fishing researchers actively search for publishable results in a large data pond. Harking, or hypothesizing after results are known, is a related strategy but describes the formulation of a theory and a research design after the results have been collected. Many, including Uri Simonsohn, believe that preregistration solves this problem. But harking can be turned into parking with ease: preregistering after the results are known. There is an infinite regress: Cheaters can always go one step back. Harking, parking, and fishing will thus always remain undetectable. These are simply inductive research strategies that researchers write up as if their research were deductive. These problems, thus, still need to be solved.

12 And One Step Back?

> The future is unwritten.
> —Joe Strummer

Science is not broken. But neither can science turn its back on the challenges it faces and ignore the problems. The crisis of confidence in science is real and deserved. The scientific community is ill-advised to neglect the crisis and its consequences, as most scientists have done hitherto. The measures taken in the past have not worked, at least not well enough.

The issue goes well beyond data fabrication and manipulation. The much deeper problem is that scientists continue to benefit from tweaking research designs and model specifications to bring empirical results into line with theoretical expectations and predictions and with ideological priors. Lee McIntyre (2019: 97) can be optimistic about the "pitch-perfect response of the scientific community" because he does not regard tweaking as fraudulent, only data fabrication and manipulation. In fact, McIntyre (2019: 95) adopts a laissez-faire approach to the manipulation of empirical results: "Clearly, p-hacking is a challenge to the thesis that scientists embrace the scientific attitude. But unless one is willing to discard a large proportion of unreproducible studies as unscientific, what other choice do we have?" In other words, he seems to suggest that we need to take potentially manipulated, unreplicable empirical results seriously because there are so many of them.

And Christie Aschwanden, using another line of defense, suggests that "scientists who fiddle around [with data to get desired empirical results] . . . aren't usually committing fraud, nor are they intending to. They're just falling prey to natural human biases that lead them to tip the scales and set up studies to produce false-positive results."[1] According to Aschwanden, if scientists estimate 100 models and include only the three in their manuscript that support their theory, then this is not fraud and unintentional. These scientists are merely unable to accept that their theory is wrong or, at least, not sufficiently supported by the empirical evidence. Well, maybe, but then they are not scientists.[2]

Harris (2017) is hesitant to call it fraud when researchers adjust models or designs in response to their results. He cites a famous quote from Richard Feynman, delivered at a Caltech commencement speech on June 14, 1974,[3] that "the first principle is that you must not fool yourself." Harris is interested in the causes that bring about false empirical results, one of which is, obviously, scientific fraud. Others include honest mistake, sloppiness, and confirmation bias. Science is difficult, and thus scientists make mistakes, take shortcuts, or do not go the extra mile it takes to come up with reliable empirical results.

Though he admits that "outright fraud also creeps into science," Harris stresses that "fraud is not a major factor" underlying the reproducibility crisis. Indeed, data fabrication and manipulation or image manipulation do not appear to be the major causes of the sensitivity of empirical results. But is it really sloppiness or an innocent mistake to use a different operationalization of a variable when results are not statistically significant? Is it merely sloppiness or an innocent mistake to drop some cases from the sample when results are not statistically significant? Is it sloppiness to drop some control variables if the results are not statistically significant? Like so many others, Harris cites survey research suggesting that 14 percent of scientists have observed fabrication and falsification, but 72 percent have observed other forms of questionable or detrimental behavior (Harris 2017: 187).

Of course, if one does not accept tweaks as fraudulent, then scientific fraud is rare. If one included tweaks, scientific fraud is no longer rare. We do not believe that model selection based on results happens in the researcher's subconscious. We believe that model selection based on results is a conscious act. Researchers who actively search for "what works" suffer from confirmation bias. And they know what they are doing.

Tweakers contribute as much as fraudsters who fabricate and manipulate data, if not more, to the lack of trust in science. The incentive to ignore this more subtle type of systematic and intentional doctoring of empirical results looms large over academia. Giving in to this incentive would be a mistake. Tweakers manipulate empirical results with intention, maybe because they believe more in their priors than in the empirical evidence that is right in front of them, maybe because they are young and need the money or old and need the fame.

The future of tweaking, the future of scientific fraud is unwritten. Maybe, just maybe, things are getting worse. Artificial intelligence is not only able to write text; AI is already able to estimate models, produce graphs, use random forest techniques to search the model space for analyses "that work". The options are endless. There can be little doubt that AI will make it easier for fraudsters to tweak their results.

At the same time, AI will improve counter-tweaking strategies. Just as tweakers can use intelligent search algorithms to find the model specification that gives the desired results, fraud chasers can employ AI in their forensic work to detect fraud, for example, by estimating a large number of models that satisfy predefined conditions.

Here is what the one and only expert on artificial intelligence has to say about this issue: "On one hand, AI technologies can be used to detect and prevent fraud by analyzing large datasets for anomalies, inconsistencies, or patterns indicative of fraudulent behavior. These systems can help identify instances of data manipulation, plagiarism, or other unethical practices in scientific research. However, on the other hand, AI itself can also be exploited to perpetrate fraud. For instance, researchers could misuse AI algorithms to generate fake data or manipulate results in a way that is difficult to detect. Moreover, AI can be used to automate the process of creating fraudulent research papers or to generate convincing but false conclusions" (ChatGPT, February 26, 2024). ChatGPT himself (we asked and he claimed that he prefers to be called "he" though he, as algorithm, has no obvious gender and, yes, we had that conversation in German), mentions six main ways in which AI can be helpful in detecting scientific fraud: by the detection of anomalies in data and the detection of plagiarized text in research papers, grant proposals, or academic articles; by analyzing and detecting citation and co-authorship cartels; by identifying manipulated or duplicated images and figures; by predicting the occurrence of fraud; and by computing the likelihood of fraudulent activity.

Tweaking is not much different from doping. Dopers and anti-doping agencies are intertwined in an arms race in which in the short run the dopers have a competitive edge but in the long run new tests emerge that are able to show that a seemingly clean cyclist has relied on disallowed "improvements" that for many years could not have been detected.

The future is unwritten. And that certainly holds for the arms race between tweakers and anti-tweakers, between fraudsters and fraud chasers. It seems to be certain, however, that AI will have its role on both sides.

Notes

Part I

1. It is not Einstein's most iconic photograph; that honor goes to Arthur Sasse's photograph of Einstein sticking out his tongue.

2. The earliest article we could find on a confidence crisis in science is by Clyve Nunn, published in 1977 (Nunn 1977). Nunn concludes that "ambivalence, not rejection, best describes public attitudes. Therefore, the scientific community should, where possible, increasingly relate science to the problems of health, the environment, energy, and society in order to maintain and expand the public's interest in science" (995). Science is there to entertain the public—a very modern theme.

It is interesting to note that, like Nunn, virtually all contemporary articles on this crisis use a question mark in the title, for example, "A Crisis of Confidence?" (Feine and Jakubovics 2021)—as if scientists are not confident enough to postulate that science is in a crisis of confidence.

3. See chapter 3 for a broader discussion of attempts to quantify scientific fraud.

4. See retractionwatch.com.

5. We use cases as illustrative examples. For a more comprehensive overview of cases of data fabrication and manipulation, see Ben-Yehuda and Oliver-Lumerman (2017).

6. semafor.com/article/04/07/2023/science-is-broken.

7. gizmodo.com/is-science-broken-major-new-report-outlines-problems-i-1834589928.

8. fivethirtyeight.com/features/science-isnt-broken/.

9. progressivescience.org/index.php/2015/08/26/science-isnt-broken-or-is-it.

10. slate.com/technology/2017/08/science-is-not-self-correcting-science-is-broken.html.

11. "One element is absolutely critical to the success of our mission to improve the human condition: trust. It's a foundational element of any relationship, but for the mutual benefit of the scientific enterprise and the people who support it, trust is essential. Simply put, a scientific endeavor that is not trusted by the public cannot adequately contribute to society and will be diminished as a result." pewtrusts.org/en/trend/archive/winter-2021/why-we-must-rebuild-trust-in-science.

12. economist.com/leaders/2013/10/21/how-science-goes-wrong.

13. *The Guardian* cites Arturo Casadevall, who has published extensively on scientific fraud and retractions (Steen et al. 2016). Casadevall suggests not only that scientific fraud is a minor problem for science, but also that what we call fraud is only a small fraction of manipulated empirical results in science. In other words, we talk too much about fraud and not enough about tweaks. theguardian.com/science/2017/feb/01/high-tech-war-on-science.

14. The main argument Aschwanden makes in the *FiveThirtyEight* article is that science is, first and foremost, hard. Who would not agree? But the fact that something is difficult is no excuse for bending the rules. Sure, it would be easier to get publishable results if we all just fabricated data we need. But that is fraud, not science. fivethirtyeight.com/features/science-isnt-broken/.

15. fivethirtyeight.com/features/science-isnt-broken/.

16. Bias and hubris in science do not just motivate scientists to manipulate empirical results in order to bring them in line with their theoretical or ideological priors. Bias also lets scientists ignore newly arising data and empirical results that contradict their priors. fivethirtyeight.com/features/science-isnt-broken/.

17. theguardian.com/science/2017/feb/01/high-tech-war-on-science.

Chapter 1

1. There are empirical studies that have found support for Lamarck's idea of the inheritance of acquired characteristics, but the vast majority of these supportive results have failed to be replicated in subsequent studies. See Wikipedia, "Paul Kammerer," en.wikipedia.org/wiki/Paul_Kammerer.

2. The sentence Macchiarini received from a Swedish court was generally interpreted as a "mild slap on the wrist." See Wikipedia, "Paolo Macchiarini," en.wikipedia.org/wiki/Paolo_Macchiarini.

3. theguardian.com/commentisfree/2023/aug/09/scientific-misconduct-retraction-watch.

4. One should be skeptical when Diederik Stapel agrees: "There are unwritten rules about which dubious practices are allowable and which aren't, even though both kinds ultimately lead to the same distortion of the results" (Stapel 2014).

5. The term "false positive" requires a binary understanding and interpretation of statistics. A false positive error accepts the hypothesis that predicts the existence of an effect when there is no such effect. Conversely, a false-negative error rejects the hypothesis that predicts the existence of an effect when the predicted effect exists.

6. The majority of researchers publishing on data manipulation and p-hacking suggest that the problem of fraud is the increase in the number of false positives. False positives are one of potentially many outcomes of scientific fraud, but when fraudsters falsify data to support a correct hypothesis, it is still fraud.

7. investopedia.com/terms/f/fraud.asp.

8. J. Achenbach, "Diederik Stapel: The lying Dutchman," *Washington Post*, November 1, 2011, washingtonpost.com/blogs/achenblog/post/diederik-stapel-the-lying-dutchman/2011/11/01/gIQA86XOdM_blog.html.

9. The article on the *Deutschlandfunk* website discusses the long-term consequences for Stapel, which included depression and serious problems finding a new job. deutschlandfunk.de/ergebnisse-um-jeden-preis-wenn-forscher-betruegen-100.html.

10. Not reporting estimation results because the model on which the results rest is misspecified is not considered fraudulent.

11. And not just among scientists: "Science is a communal enterprise built on trust. Referees and editors generally take data at face value and assume that the authors have honestly reported and analyzed their results. Reviewers are asked to judge whether a report's conclusions are solid based on the data and not whether the data themselves are fraudulent. The system is not set up to work any other way; if editors and referees distrusted all authors and assumed that every result was potentially fake, few papers would be published. There are not enough known cases of outright fraud to justify the expense (in time, money or experimental animals) to require that all data be duplicated in independent laboratories" (*Nature Neuroscience* 2006).

Chapter 2

1. The arguably more common opposite strategy was chosen by Francesca Gino: Deny fraud, blame research students if some data errors are undeniable, talk to no one, and sue your university and the whistleblowers who exposed your misdeeds.

2. nytimes.com/2013/04/28/magazine/diederik-stapels-audacious-academic-fraud.html.

3. See Wikipedia, "Diederik Stapel," en.wikipedia.org/wiki/Diederik_Stapel. Stapel may have been the first social psychologist to be caught fabricating and manipulating data. But other cases of authors who had their articles retracted would follow, including those of Jens Förster, Lorenza Colzato, Francesca Gino, and Dirk Smeesters, to name just a few that we mention in this book.

4. This is known as Edsel Murphy's general law.

5. news.ki.se/news-archive/the-macchiarini-case-timeline.

6. nytimes.com/2014/11/25/world/leading-surgeon-is-accused-of-misconduct-in-experimental-transplant-operations.html.

7. Ibid.

8. thelancet.com/journals/lancet/article/PIIS0140-6736(15)00118-X/fulltext.

9. Nevertheless, Gross (2016: 700) believes that a typical fraudulent researcher can be described as follows: "This scientist is a bright and ambitious young man working in an elite institution in a rapidly moving and highly competitive branch of modern biology or medicine, where results have important theoretical, clinical, or financial implications. He has been mentored and supported by a senior and respected establishment figure who is often the co-author of many of his papers but may have not been closely involved in the research."

10. Wikipedia, "Diederik Stapel," en.wikipedia.org/wiki/Diederik_Stapel.

11. nytimes.com/2015/05/23/opinion/whats-behind-big-science-frauds.html.

12. spektrum.de/magazin/die-affaere-stapel/1140956.

13. retractionwatch.com/2015/05/20/author-retracts-study-of-changing-minds-on-same-sex-marriage-after-colleague-admits-data-were-faked.

14. retractionwatch.com.

15. skysports.com/more-sports/cycling/news/15264/8160021/lance-armstrongs-success-prompted-jan-ullrich-to-use-drugs-again. We occasionally see a very similar argument from detected fraudsters who claim to be mere victims of previous fraudsters and that they simply had to level the playing field with them.

Chapter 3

1. Lichtenthaler continued to publish after leaving the University of Mannheim in 2015. Between 2017 and 2019, four articles appeared in B-journals. Between 2020 and 2023, he published another fourteen articles. He continues to be cited. retractionwatch.com/category/by-author/ulrich-lichtenthaler/.

2. smh.com.au/world/when-research-findings-explode-20120919-2674r.html.

3. Other studies across a variety of fields and disciplines include (but are not limited to) Camerer et al. (2016), Chang and Li (2017), Laitin and Reich (2017), Camerer et al. (2018), Altmejd et al. (2019), Kelly (2019), and Fraser et al. (2020).

4. According to the *Science* article (Bohannon 2015), the authors of the replication study express surprise that replications are not more common. However, their efforts

highlight the fact that systematic replication is costly and that external funding for replication studies is virtually nonexistent. This leaves researchers with the difficult choice of conducting research on an original contribution or analyzing whether someone else's contributions are valid.

5. Imagine that empirical researchers have to test 1,000 hypotheses in a field where the vast majority of hypotheses are very unlikely to be true. Specifically, suppose that only 100 out of 1,000 hypotheses tested are actually true. If these hypotheses are tested honestly, using a conventional threshold of statistical significance of 0.05, $0.05 \times 900 = 45$ tests will find the effect to be statistically significant if in fact there is no effect. Researchers call this "false positives." Now combine this with a relatively low power of 40 percent. This means that empirical studies will correctly identify only 40 percent of the 100 hypotheses that are actually true; that gives forty true positives in 1,000 analyses. Since researchers tend to publish only statistically significant results, the majority of published research is false.

6. An increasing number of journals have stopped reporting p-values and instead report effect sizes. We believe that reporting effect sizes with confidence intervals around them is superior to reporting coefficients and p-values. It should be noted, however, that effect sizes can also be fabricated and manipulated.

Chapter 4

1. The last person ranked on the leaderboard is Prashant Sharma, who still had twenty-eight articles retracted as of August 2023—bad enough for thirtieth place on the leaderboard. retractionwatch.com/the-retraction-watch-leaderboard/.

2. theatlantic.com/national/archive/2011/10/is-your-kindly-neighbor-a-serial-killer/247290.

3. Although Lieber has not retracted any of his articles, citations of his work started to decline in 2011, a decade before his conviction. The decline in citations has continued since then, but it has not accelerated. Wikipedia, "Charles M. Lieber," en.wikipedia.org/wiki/Charles_M._Lieber.

4. retractionwatch.com/2021/05/20/the-notices-are-utterly-unhelpful-a-look-at-how-journals-have-handled-allegations-about-hundreds-of-papers/.

5. Three whistleblowers describe the psychological rollercoaster they went through: "You start feeling isolated and a bit like a cheater because there are important things going on, affecting also your colleagues, yet you cannot talk about it. You have to trust your superiors and the scientific integrity committee even though you don't really trust anyone. You feel stuck; everyone around you moves forward but you cannot envision a future, or even what you would want the future to look like. There's a constant struggle between knowing you did the right thing and the fear

of losing everything." retractionwatch.com/2019/12/11/a-long-and-lonely-process-whistleblowers-in-a-misconduct-case-speak-out.

6. retractionwatch.com/2018/07/30/how-institutions-gaslight-whistleblowers-and-what-can-be-done.

7. See also Wikipedia, "David Sanders (biologist)," wikipedia.org/wiki/David_Sanders_(biologist).

8. thecut.com/2015/05/co-author-of-the-faked-study-speaks-out.html. The retraction of the LaCour and Green *Science* article is a good case to discuss whether the co-authors are the first victims of fraudsters or guilty as well. There are strong arguments for both views. On the one hand, co-authors were insufficiently involved in the research process if they did not notice that their co-author fabricated or manipulated data. On the other hand, they were not the ones who fabricated or manipulated data. retractionwatch.com/2021/07/01/a-scientist-critic-was-sued-and-won-but-did-not-emerge-unscathed-this-is-his-story.

9. scribd.com/document/554975819/The-Case-of-Diederik-Stapel.

10. Biagioli suggests that fraudsters invent co-authors for three different reasons: (1) to divert attention and suspicion from the fact that fraudsters publish "too much," (2) to hide the fact that papers require skills that fraudsters do not have, and (3) to associate their papers with a more prestigious university.

The story of invented co-authors takes a bizarre turn when even the main author may not exist and works for an unaccredited university. Alireza Heidari is such a case. He is the proud author of a review article that has fewer words than self-citations, but it remains unclear whether he is just the pseudonym of a paper mill or a real person. He claims to work for California South University, which has a website that shows pictures of tall university buildings that do not seem to exist in Irvine, and which uses 14731 Comet Street, a small house in a residential neighborhood, as its mailing address, a house that is pixilated in Google Street View. A predatory researcher working for a predatory university?

11. chronicle.com/article/the-unraveling-of-michael-lacour.

12. Andrew Gelman suggests that there is more than one reason to be suspicious of the Beall and Tracy article. However, none of these reasons prove the manipulation of empirical results. statmodeling.stat.columbia.edu/2013/07/31/response-by-jessica-tracy-and-alec-beall-to-my-criticism-of-their-paper/.

13. replicationindex.com/2014/12/24/women-in-red.

14. Gelman argues that tweaking is a gray area of manipulating empirical results. Many, certainly too many, authors do it, few tweakers get caught, and, in the end, science begins to steer directly into a replication crisis. statmodeling.stat.columbia.edu/2013/07/31/response-by-jessica-tracy-and-alec-beall-to-my-criticism-of-their-paper/.

15. purdue.edu/uns/x/2008b/080827WoodsonTaleyarkhan.html.

16. fivethirtyeight.com/features/how-two-grad-students-uncovered-michael-lacour-fraud-and-a-way-to-change-opinions-on-transgender-rights/.

17. livescience.com/34737-fertile-peak-women-wear-red.html.

18. discovermagazine.com/health/women-are-more-likely-to-wear-red-or-pink-at-peak-fertility.

19. time.com/2813381/hurricanes-female-names.

20. DataColada, "[98] Evidence of Fraud in an Influential Field Experiment About Dishonesty," DataColada.org/98.

21. DataColada.org, posts 109–112.

22. Gino is not just an ordinary researcher. She was the Tandon Family Professor of Business Administration at Harvard Business School, one of the highest-paid researchers at one of the highest-paid research universities in the world, and has been named one of the world's top forty business professors under forty. Her research has been featured in *The Economist*, *The New York Times*, *Newsweek*, *Scientific American*, and *The Wall Street Journal*, among others.

23. DataColada.org/. An interesting aspect of this story is that Simonsohn and coauthors informed Harvard University of their findings and gave the university eighteen months to conduct its own evaluation of misconduct in the work of Francesca Gino. For interesting background and an assessment of the process, see Pete Judo's excellent YouTube coverage of the case: Pete Judo, "Academia Is BROKEN! Harvard Fake Data Scandal Explained," YouTube, June 29, 2023, youtube.com/watch?v=d2Tm3Yx4HWI.

24. DataColada, "[109] Data Falsificada (Part 1): Clusterfake," DataColada.org/109.

25. Our graph uses a slightly different design, but clearly adopts the logic of a figure used by Simonsohn et al. on their DataColada blog.

26. DataColada, "[114] Exhibits 3, 4, and 5," DataColada.org/114.

27. Francesca v Harvard.org, "Breaking My Silence," francesca-v-harvard.org.

28. https://www.lesswrong.com/posts/cGLP5iEidrJQ7JhDJ/linkpost-francesca-v-harvard.

29. DataColada, "[110] Data Falsificada (Part 2): 'My Class Year Is Harvard,'" DataColada.org/110.

30. DataColada, "[114] Exhibits 3, 4, and 5."

31. DataColada, "[111] Data Falsificada (Part 3): 'The Cheaters Are Out of Order,'" DataColada.org/111.

32. DataColada, "[112] Data Falsificada (Part 4): 'Forgetting the Words,'" DataColada.org/112.

33. DataColada, "[114] Exhibits 3, 4, and 5."

34. DataColada, "Exhibit 5," DataColada.org/wp-content/uploads/Harvard-Report-on-Gino.pdf.

35. theguardian.com/science/2012/sep/13/scientific-research-fraud-bad-practice.

36. Ibid.

37. retractionwatch.com/2019/04/23/a-new-form-of-plagiarism-when-researcher-fake-co-authors-names.

38. medium.com/@brainosoph/the-ghost-of-stronzo-bestiale-and-other-fake-scientific-authors-b4e1c6b5f424.

39. nytimes.com/2013/04/28/magazine/diederik-stapels-audacious-academic-fraud.html.

Chapter 5

1. This applies to frequentist statistical testing. We have no doubt that tweakers can similarly manipulate Bayesian statistical testing, and Simonsohn (2014) provides evidence in this regard.

2. Evidence suggests that p-hacking is common (Gelman and Loken 2014; Head et al. 2015; Open Science Collaboration 2015), but this evidence is at the aggregate level in the form of bunching of p-values just below a cutoff for statistical significance across a number of papers in a discipline, rather than identifying individual studies. P-hacking at the level of a single study is difficult to detect unless researchers resort to highly implausible modeling choices due to a lack of talent. When researchers instead use tweaks that are indistinguishable from honest specification searches, proving malicious intent becomes impossible. Functional form assumptions, dropping or adding seemingly irrelevant control variables, specification of the estimator, and standard errors are all examples of specification choices that can have a large impact on relevant quantities of interest—but are indistinguishable from specification choices that researchers must make anyway.

3. Many scientists seem to accept the p-value as the criterion that divides results into true, usually $p \leq 0.05$, and false, $p > 0.05$. Sometimes the p-value is understood as the probability that an empirical result is caused by random processes. This is consistent with Ronald Fisher's proposal of the p-value as an "exact test" of statistical significance. As for urban legends, Fisher based his test on the story of Muriel Bristol, also known as the tea-tasting lady. Muriel Bristol claimed that she could tell the order in which tea and milk were poured into a cup by tasting the resulting

brew. Since there are two alternative orders, tea first and milk first, guessing would have a probability of being correct 50 percent of the time. However, the probability of getting 2 out of 2 trials correct by chance is 25 percent, the probability of getting 3 out of 3 correct is 12.5 percent, and so on. The probability that the lady got 9 trials right and 1 wrong by chance is a little less than 1 percent. Fisher called this probability the p-value. In regression analysis, the p-value measures a different quantity. It quantifies the probability of finding a result that is equal to or greater than an estimated positive coefficient, or equal to or less than an estimated negative coefficient, under the assumption that the null hypothesis being tested is true. This conditioning assumption immediately destroys any hope that the p-value provides any information about the probability that an effect or hypothesis is true. As Christie Aschwanden, former senior science writer at *FiveThirtyEight*, puts it, "we want to know if results are right, but a p-value doesn't measure that. It can't tell you the magnitude of an effect, the strength of the evidence or the probability that the finding was the result of chance." And it gets worse: The estimation model must be correctly specified so that the estimated point estimate is correct and errors must be drawn from a perfect normal distribution, which can only be taken for granted if the sample size approaches infinity. Contrary to the tea tasting example, the p-value in regression analysis is not a fixed quantity that can be calculated in advance. The p-value is an estimated quantity, derived from a particular estimation model and therefore dependent on its specification. In fact, both the coefficient and the standard error from which the p-value is derived are estimates. fivethirtyeight.com /features/not-even-scientists-can-easily-explain-p-values.

4. fivethirtyeight.com/features/not-even-scientists-can-easily-explain-p-values.

5. The term "p-hacking" has been traced back to Simonsohn, Nelson, and Simmons during a psychology conference. The term, though easily memorizable, has triggered a debate among methodologists: "Critics argue that the word 'hacking' implies it refers only to intentionally deceptive manipulation. In actuality, the word applies to both unintentional and intentional cases." statisticsbyjim.com /hypothesis-testing/p-hacking.

6. Chapter 7 discusses a regression discontinuity analysis that seems to have been specified to give the largest effect size in symmetric bands.

7. "As an unintended consequence, researchers may now anticipate this selection and consider that it is a stumbling block for their ideas to be considered. As such, among a set of acceptable specifications for a test, they may be tempted to keep those with the highest statistics in order to improve their chances of being published. Keeping only such specifications would lead to inflation in the statistics of observed tests" (Brodeur et al. 2016).

8. In statistics, this chance is the random deviation of random draw from a standard normal distribution from the assumed standard normal distribution.

9. Piketty's *Capital in the Twenty-First Century* (2014) sparked a similar controversy. According to Giles, Piketty's work contains a number of errors that have the potential to undermine crucial arguments in his book. If nothing else, the Piketty controversy demonstrates that fraudulent data manipulation can often be difficult to separate from the data manipulation required when researchers attempt to synthesize information from disparate sources. ft.com/content/e1f343ca-e281-11e3-89fd-00144feabdc0.

10. "If I'd designed a study for six groups of participants who'd all been given slightly different tasks or shown different things and one group gave a strange result that I couldn't explain, I would change the design so that the strange result disappeared, but the overall design still appeared coherent. Then it became a three- or four- or five-group design, instead of six. . . . If an experiment didn't work, I would go hunting in the dataset for outliers . . . and look for reasons why it was OK to throw them out. If the participants were a little older, or younger, or slower, or faster, or in whichever way not quite 'normal,' then maybe it was OK to eliminate them, to bring the results more in line with what I'd been expecting. I knew that this was at the outer limits of what was acceptable, or even beyond those limits in some cases, but . . . I wanted the results so badly" (Stapel 2014).

11. fastcompany.com/1683190/economist-steven-levitt-on-why-data-needs-stories. Levitt argues that data need stories. That claim might be correct. But not just any story.

12. For evidence that past disaster experience reduces current disaster mortality, see our work on tsunami mortality (Plümper et al. 2017).

13. Results (or data and do-file) are available at doi.org/10.7910/DVN/SIXM7E. We include all models in the do-file, which allows replication of all results presented in this book.

14. The story of the Camille hurricane party may be an urban legend. It is said that of the twenty-four people who attended the party, all but one died telling the story to the media. As we know now, the truth is (or seems to be) that eight people died in the Richelieu apartments during Camille, and none of these attended a party. What killed them was the collapsing buildings and the fact that they did not evacuate inland. Still, hurricane parties do happen (Hearn 2009).

Chapter 6

1. Cornell University refused to publish the report that led to this conclusion. wrvo.org/2018-11-09/cornell-finds-laboratory-head-falsified-data.

2. retractionwatch.com/2017/02/02/backlash-prompts-prominent-nutrition-researcher-reanalyze-multiple-papers.

3. statmodeling.stat.columbia.edu/2017/01/25/statistical-heartburn-attempt-digest-four-pizza-publications-cornell-food-brand-lab/. Andrew Gelman's blog post offers

some insights into the motives of fraudsters: "I get the impression that Diederik Stapel, Jens Foerster, and Bruno Frey knew exactly what they were doing: they were gaming the system. Given Frey's writing about how the system was too much like a game, he in particular may have felt justified—perhaps he felt that everyone else was doing it, so he should too. Moving to someone like Marc Hauser: I expect that he knew exactly what he was doing too. Yes, he was cheating, but I'm guessing he felt that he was cheating in support of a larger truth. In his mind, he knew what was scientifically correct and he didn't want the messiness of real data to stand in the way or to distract people from the underlying truth. But Wansink . . . he's kind of amazing. He just seems like a pure careerist, having 100% bought into the goal that the purpose of the scientific endeavor is: publications, grants, fame, successful students, etc. Based on his blog post and his responses to comments, he seems genuinely interested in making people happy. The quality of the work doesn't matter at all."

4. It's worth mentioning that some fields allow data collection until results are definitive. If a medical drug shows promising benefits before the planned number of cases has been reached, randomized trials are often stopped. The thinking is that it's unethical to deny participants in the control group the benefit of the treatment. Similarly, randomized trials may be stopped if serious side effects are found in the treatment group.

5. "The publication of this article rocked psychology and triggered a credibility crisis in psychological science. Unforeseen by Bem, the article did not sway psychologists to believe in time-reversed causality. Rather, it made them doubt other published findings in psychology." replicationindex.com/2018/01/05/bem-retraction.

6. According to Bem, going on a fishing expedition is not immoral. psychology.yale.edu/sites/default/files/bemempirical.pdf.

7. washingtonpost.com/national/health-science/the-press-release-crime-of-a-biotech-ceo-and-its-impact-on-scientific-research/2013/09/23/9b4a1a32-007a-11e3-9a3e-916de805f65d_story.html.

8. Ibid.

9. Ibid.

10. Ibid.

11. We are grateful to Raymond Duch from Oxford University for allowing us to use his unpublished experimental data for illustrative purposes in this chapter.

Chapter 7

1. It could be that in real life, most discontinuity thresholds are rather fuzzy, meaning that not everyone who qualifies for treatment according to the forcing variable is guaranteed to receive treatment, and not everyone who falls below the threshold

is left out. With a sharp discontinuity threshold, selection into treatment is deterministic (Cattaneo and Titiunik 2022); with a fuzzy discontinuity, selection is probabilistic. For instance, there are means-tested eligibility thresholds that may stop individuals who meet the threshold from accessing welfare benefits if these are in limited supply. Meanwhile, some who do not meet the threshold find other ways to benefit. Universities have a limited number of places and so reject some applicants who meet the minimum entry grade requirements but admit some others who fall below the threshold for other reasons.

2. For instance, Urquiola and Verhoogen (2009) think that the threshold isn't exogenous and that it's actually strategically manipulated by individuals for class sizes. Similarly, Barreca et al. (2011) looked at birth weight, and Caughey and Sekhon (2011) looked at the closeness of elections.

3. Caughey and Sekhon (2011: 405) argue that the continuity of potential outcomes assumption is not satisfied in the case of "narrow electoral" races: "Excessive reliance on parametric extrapolation caused him and other scholars to miss the covariate imbalance, which worsens dramatically in the closest elections."

4. In an ideal scenario, the force in question is nature, which would lead to the claim of a "natural experiment." In instances where nature does not oblige, the force in so-called quasi-natural experiments is human, but decisions are made at a higher aggregate level, which means that the cases themselves do not have the discretion to opt into or out of treatment.

5. All causal models lose some information along the way to get "unbiased and well-identified estimates of the true causal effect." Thus, these research designs come at the price of a decrease in efficiency—a reduction in the signal-to-noise ratio and, as a consequence, an increase in the influence of noise on estimation results.

6. All "spatial instruments" are ones where researchers use weighted regional or global averages of the variable to be instrumented for. These have been shown to be pretty unreliable (Betz et al. 2018), but the issue goes well beyond that. It includes, for example, the much-misused colonial settler mortality and rainfall instruments (Mellon 2025).

In their review of the accounting field, Larcker and Rusticus (2010: 187) come to a striking conclusion: "In general, the variables selected as instruments seem largely arbitrary and are not justified by any rigorous theoretical discussion." Paul R. Rosenbaum finds that "in reading the scientific literature, many claimed instruments seem implausible as instruments" (Rosenbaum 2010: 139). If he is right that instruments "are rare, but useful when they exist" (140), then how is it that Google Scholar finds some 319,000 results for the exact term "instrumental variable" combined with "result" or "results" (accessed August 16, 2024)?

7. First, they estimate a regression discontinuity model of overrepresentation of published p-values slightly below the 0.05 threshold. This is a commonly used indicator of p-hacking, which assumes that the likelihood density function of p-values is "continuous and infinitely differentiable" (Brodeur et al. 2020: 3635). A larger than expected number of coefficients that just reach statistical significance indicates that researchers have pushed the level of significance above the threshold that separates publishable from unpublishable manuscripts. Second, and following Gerber and Malhotra (2008), they conduct caliper tests that also focus on the overrepresentation of p-values slightly below the threshold. And third, they compare the observed distribution of test statistics "to a counterfactual distribution that we would expect in the absence of p-hacking and publication bias" (Brodeur et al. 2020: 3636).

8. See Open ICPSR, "Replication Data For: Buy the Book? Evidence on the Effect of Textbook Funding on School-Level Achievement," principle investigator Kristian L. Holden, www.openicpsr.org/openicpsr/project/113641/version/V1/view.

Part III

1. westyorkshire.police.uk/advice/10-principles-crime-prevention/10-principles-crime-prevention/10-principles-crime-prevention.

Chapter 8

1. pewtrusts.org/en/trend/archive/winter-2021/why-we-must-rebuild-trust-in-science.

2. zenodo.org/record/6472827#.Y9k_i3bMKUk. The Deutsche Forschungsgemeinschaft (2022) sets rules but doesn't say how they should be enforced. It makes you wonder what it would do if universities decided that the Deutsche Forschungsgemeinschaft was responsible for making sure that research it funded was done properly and followed good practice.

3. The replication study by the Open Science Collaboration is a rare example that proves the rule: The significant attention its authors and contributors received came from two main factors. First, it was a huge study, and second, it was the first study of its kind, replicating a very large number of studies simultaneously and collaboratively. Later studies that replicated other studies in different fields got much less attention. In social psychology, which has a lot of fraud, the number of times articles that couldn't be replicated were cited went down. But only by about 5 percent, and only 3 percent of the articles that cited the article that couldn't be replicated also cited the replication study (von Hippel 2022).

4. "Throughout the entire process, until the very last moment when multiple 'smoking guns' finally appeared, Broockman was consistently told by friends and

advisers to keep quiet about his concerns lest he earn a reputation as a troublemaker, or, worse, someone who merely replicates and investigates others' research rather than plant a flag of his own." vox.com/2015/6/3/8720975/science-fraud-replication.

5. chicagotribune.com/news/ct-xpm-1989-08-16-8901050101-story.html.

6. onlineethics.org/sites/onlineethics/files/2021-04/Ethics%20in%20the%20Science%20Classroom_0.pdf.

7. discover.hubpages.com/education/The-High-Profile-Cases-of-Research-Misconduct.

8. chicagotribune.com/news/ct-xpm-1989-08-16-8901050101-story.html.

9. retractionwatch.com/2011/09/27/savaskan-group-forced-to-retract-faseb-j-paper-corrects-cell-study-with-duplicated-figure.

10. retractionwatch.com/2013/06/04/bitter-legal-fight-leads-to-a-retracted-retraction.

11. retractionwatch.com/2024/05/17/how-the-karolinska-protected-paolo-macchiarini-and-whistleblowers-paid-the-price.

12. See eugeniereichlaw.com/.

13. drgeoffnutrition.wordpress.com/2017/06/24/Jatinder-ahluwalia-a-former-colleague-accused-of-multiple-acts-of-research-fraud.

14. ucl.ac.uk/governance-compliance/sites/governance_compliance/files/research-misconduct-panel.pdf.

15. timeshighereducation.com/news/imperial-releases-ahluwalia-report/2006095.article.

16. retractionwatch.com/category/by-author/Jatinder-ahluwalia. The evaluation of possible scientific misconduct was made difficult by the reluctance of the industrial collaborator on Dr. Ahluwalia's studentship to grant access to his lab books.

17. However, it is notable that laboratory assistants frequently lack permanent positions, whereas tenured professors typically possess them. Consequently, the impact of position within the academic hierarchy on the prevalence of scientific fraud cannot be deduced from disparate survival rates alone.

18. faz.net/aktuell/feuilleton/debatten/urteil-zu-wissenschaftsfreiheit-im-fall-guerot-die-kehrseite-des-grundrechts-19945576.html.

19. theconversation.com/tougher-action-needed-in-the-fight-against-scientific-fraud-68076.

20. newscientist.com/article/mg22329864-100-its-time-to-criminalise-serious-scientific-misconduct.

21. chris-said.io/2024/06/17/the-case-for-criminalizing-scientific-misconduct.

22. masslive.com/news/2010/01/dr_scott_reuben_former_chief_o.html.

23. masslive.com/news/2010/06/scott_reuben_a_former_baystate.html?ref=us.

24. abc.net.au/news/2013-11-07/uq-investigates-fresh-claims-of-research-mis conduct/5077276.

25. retractionwatch.com/2016/10/24/Australian-court-finds-parkinsons-researcher -guilty-of-fraud.

26. the-scientist.com/nutrition-retracts-2001-paper-49012.

27. cbc.ca/news/canada/newfoundland-labrador/did-nothing-to-harm-chandra -researcher-s-reputation-mun-1.633358.

28. YouTube.com/watch?v=gKYr1VerT4A; part 2: YouTube.com/watch?v=2R7-QM8e KCM; part 3: YouTube.com/watch?--v=YbQNlm7sw5A.

29. www.cbc.ca/news/canada/newfoundland-labrador/chandra-paying-legal-fees-1 .3322284.

Chapter 9

1. Benjamin et al. (2018) try to tackle this issue by suggesting that results that would have passed the old threshold of 0.05, but fall short of the new threshold of 0.005, can be considered "suggestive" and should still be considered publishable—as if authors, reviewers, and editors were satisfied with "suggestive." If authors, reviewers, and editors behaved more like grown-ups, rather than obsessing over essentially arbitrary thresholds of significance, many of our problems would be less, well, significant. There would still be tweaking, but p-hacking would not rule supreme.

Chapter 10

1. nautil.us/how-the-biggest-fabricator-in-science-got-caught-235421.

2. University of Würzburg (2012).

3. See Embassy of Good Science, "Forensic Statistics to Detect Data Fabrication," embassy.science/wiki/Theme:467f5cf6-d41f-42a0-9b19-76556579845d.

4. Even Simonsohn, Simmons, and Nelson (2020) admit that "specification curve analysis provides a (partial) solution to the problem of selectively reported results. Readers expecting a judgement-free solution, one where researchers' viewpoints do not influence the conclusions, will be disappointed by this (and any other) solution. . . . The goal to eliminate subjectivity is unattainable (and not, in our view, desirable)."

5. We do not really know what "sufficient" means, and when, how, and if sufficient matters become insufficient. DataColada, "[104] Meaningless Means: Some Fundamental Problems with Meta-Analytic Averages," DataColada.org/104.

6. Debasmita Mandal, Pingfu Fu, and Alan D. Levine "REDOX Regulation of IL-13 Signalling in Intestinal Epithelial Cells: Usage of Alternate Pathways Mediates Distinct Gene Expression Patterns" (left image) and Debasmita Mandal and Alan D. Levine "Elevated IL-13Rα2 in Intestinal Epithelial Cells from Ulcerative Colitis or Colorectal Cancer Initiates MAPK Pathway" (right images).

7. See Bik's discussion in PubPeer for additional information: https://pubpeer.com/publications/F179024C7951219C4595163D79A87D.

8. He also reports: "In January, 2014, Bik was scrolling through a suspicious dissertation when she began glancing at the images, too. They included photographs known as Western blots, in which proteins appear as dark bands. Bik thought that she'd seen one particular protein band before—it had a fat little black dot at one end. Elsewhere in the dissertation, she found the same band flipped around and presented as if it were data from a different experiment. She kept looking, and spotted a dozen more Western blots that looked copied or subtly doctored. She learned that the thesis, written by a graduate student at Case Western Reserve University, had been published as two journal articles in 2010." newyorker.com/science/elements/how-a-sharp-eyed-scientist-became-biologys-image-detective.

9. nytimes.com/interactive/2022/10/29/opinion/science-fraud-image-manipulation-photoshop.html.

10. boardoftrustees.stanford.edu/sites/g/files/sbiybj31576/files/media/file/scientific-panel-final-report.pdf.

11. statnews.com/2022/11/30/stanford-president-altered-images/.

12. stanforddaily.com/2023/07/19/stanford-president-resigns-over-manipulated-research-will-retract-at-least-3-papers.

13. It certainly is not rare. The blog *For Better Science* has an article, written by Leonid Schneider, that has a long list of image fraud cases. forbetterscience.com/2020/07/07/science-misconduct.

14. See NASA.gov, "NASA Images," nasa.gov/multimedia/imagegallery/index.html.

Chapter 11

1. ipr.northwestern.edu/news/2024/an-existential-crisis-for-science.html.

2. fivethirtyeight.com/features/science-isnt-broken.

3. researchprofessionalnews.com/rr-news-europe-infrastructure-2024-2-elsevier-parent-reports-10-hike-in-profits-for-2023/.

4. Theoretically, and as the number of participants goes to infinity, the dependence of results on sample selection would not exist if treatments were perfectly randomized.

5. fivethirtyeight.com/features/science-isnt-broken.

Chapter 12

1. One might ask how Aschwanden could possibly know whether tweakers manipulated data to advance in academia, get a promotion and a pay raise, attract taxpayers' money (called research grants), and/or maximize citations. Or could it be that they are just innocent researchers who do their best and just happen to select models based on results because they feel that the models they select are close to the truth? The point is not that all tweakers cheat just to get their ideas published or for other self-interest. The point is that there is plenty of evidence that tweakers cheat for self-interest because replication studies show significantly smaller effect sizes and significantly wider confidence intervals. Deviations from replication results are systematic, and that would not be the case if Aschwanden was right and deviations were stochastic. fivethirtyeight.com/features/science-isnt-broken/.

2. We adopted this line from the great Johnny Cash, or at least from the screenwriter of the movie *Walk the Line,* who retorts when challenged that Christian folk do not want him singing before criminals in prison: "Well, they are not Christians then."

3. See Speakola, "Richard Feynman: 'The First Principle Is That You Must Not Fool Yourself.' Cargo-Cult Science Speech, Caltech 1974," June 14, 1974, speakola.com/grad/richard-feynman-caltech-1974.

References

Acemoglu, D., Johnson, S., and Robinson, J. A. (2001). The colonial origins of comparative development: An empirical investigation. *American Economic Review*, 91(5), 1369–1401.

Aczel, B., Szaszi, B., and Holcombe, A. O. (2021). A billion-dollar donation: Estimating the cost of researchers' time spent on peer review. *Research Integrity and Peer Review*, 6(1), 1–8.

Alcock, J. (2011). Back from the future: Parapsychology and the Bem affair. *Skeptical Inquirer*, 35(2), 31–39.

Altmejd, A., Dreber, A., Forsell, E., Huber, J., Imai, T., Johannesson, M., . . . and Camerer, C. (2019). Predicting the replicability of social science lab experiments. *PLOS One*, 14(12), e0225826.

Amrhein, V., and Greenland, S. (2018). Remove, rather than redefine, statistical significance. *Nature Human Behaviour*, 2(1), 4–4.

Angrist, J. D. (2004). Treatment effect heterogeneity in theory and practice. *Economic Journal*, 114(494), C52–C83.

Angrist, J. D., and Krueger, A. B. (2001). Instrumental variables and the search for identification: From supply and demand to natural experiments. *Journal of Economic Perspectives*, 15(4), 69–85.

Aron-Dine, A., Einav, L., and Finkelstein, A. (2013). The RAND health insurance experiment, three decades later. *Journal of Economic Perspectives*, 27(1), 197–222.

Assem, Y., Adie, S., Tang, J., and Harris, I. A. (2017). The over-representation of significant p values in abstracts compared to corresponding full texts: A systematic review of surgical randomized trials. *Contemporary Clinical Trials Communications*, 7, 194–199.

Athey, S., and Imbens, G. W. (2017). The state of applied econometrics: Causality and policy evaluation. *Journal of Economic Perspectives*, 31(2), 3–32.

Augurzky, B., and Kluve, J. (2007). Assessing the performance of matching algorithms when selection into treatment is strong. *Journal of Applied Econometrics*, 22(3), 533–557.

Bakkensen, L. A., and Larson, W. (2014). Population matters when modeling hurricane fatalities. *Proceedings of the National Academy of Sciences*, 111(50), E5331–E5332.

Barreca, A. I., Guldi, M., Lindo, J. M., and Waddell, G. R. (2011). Saving babies? Revisiting the effect of very low birth weight classification. *Quarterly Journal of Economics*, 126(4), 2117–2123.

Bartels, L. M. (1997). Specification uncertainty and model averaging. *American Journal of Political Science*, 41(2), 641–674.

Baruch, Y. (2024). HARKing can be good for science: Why, when, and how c/should we Hypothesizing After Results are Known or Proposing research questions After Results are Known. *Human Resource Management Journal*, 34(4), 865–878.

Beall, A. T., and Tracy, J. L. (2013). Women are more likely to wear red or pink at peak fertility. *Psychological Science*, 24(9), 1837–1841.

Begley, C. G., and Ellis, L. M. (2012). Raise standards for preclinical cancer research. *Nature*, 483, 531–533.

Bem, D. J. (2011). Feeling the future: Experimental evidence for anomalous retroactive influences on cognition and affect. *Journal of Personality and Social Psychology*, 100(3), 407.

Benford, F. (1938). The law of anomalous numbers. *Proceedings of the American Philosophical Society*, 78(4), 551–572.

Benjamin, D. J., and Berger, J. O. (2019). Three recommendations for improving the use of p-values. *American Statistician*, 73(suppl. 1), 186–191.

Benjamin, D. J., Berger, J. O., Johannesson, M., Nosek, B. A., Wagenmakers, E. J., Berk, R., . . . and Johnson, V. E. (2018). Redefine statistical significance. *Nature Human Behaviour*, 2(1), 6-10.

Ben-Yehuda, N., and Oliver-Lumerman, A. (2017). *Fraud and Misconduct in Research: Detection, Investigation, and Organizational Response*. University of Michigan Press.

Betz, T., Cook, S. J., and Hollenbach, F. M. (2018). On the use and abuse of spatial instruments. *Political Analysis*, 26(4), 474–479.

Bik, E. M., Casadevall, A., and Fang, F. C. (2016). The prevalence of inappropriate image duplication in biomedical research publications. *MBio*, 7(3), 10–1128.

Bishop, D. V., and Thompson, P. A. (2016). Problems in using p-curve analysis and text-mining to detect rate of p-hacking and evidential value. *PeerJ*, 4, e1715.

Blake, K. R., Dixson, B. J., O'Dean, S. M., and Denson, T. F. (2017). No compelling positive association between ovarian hormones and wearing red clothing when using multinomial analyses. *Hormones and Behavior*, 90, 129–135.

Bohannon, J. (2015). Many psychology papers fail replication test. *Science*, 349 (6251), August 28.

Bourguignon, F., Nuñez, J., and Sanchez, F. (2003). A structural model of crime and inequality in Colombia. *Journal of the European Economic Association*, 1(2–3), 440–449.

Bradshaw, C. W. (1981). A re-examination of Phillips' death-dip hypothesis. *Journal of General Psychology*.

Brainard, J. (2018). Rethinking retractions. *Science* 362 (6413): 391–395.

Brambor, T., Clark, W. R., and Golder, M. (2006). Understanding interaction models: Improving empirical analyses. *Political Analysis*, 14(1), 63–82.

Brembs, B., Huneman, P., Schönbrodt, F., Nilsonne, G., Susi, T., Siems, R., . . . and Rodriguez-Cuadrado, S. (2023). Replacing academic journals. *Royal Society Open Science*, 10(7), 230206.

Broad, W., and Wade, N. (1983). *Betrayers of the Truth: Fraud and Deceit in the Halls of Science*. Century Publishing.

Broad, W., and Wade, N. (2015). Fraud and the structure of science. In Erwin, E., Gendin, S., and Kleiman, L. (eds.), *Ethical Issues in Scientific Research* (pp. 68–90). Routledge.

Brodeur, A., Cook, N., and Heyes, A. (2020). Methods matter: P-hacking and publication bias in causal analysis in economics. *American Economic Review*, 110(11), 3634–3660.

Brodeur, A., Lé, M., Sangnier, M., and Zylberberg, Y. (2016). Star wars: The empirics strike back. *American Economic Journal: Applied Economics*, 8(1), 1–32.

Bruns, S. B., and Ioannidis, J. P. (2016). P-curve and p-hacking in observational research. *PLOS One*, 11(2), e0149144.

Callaway, E. (2011). Report finds massive fraud at Dutch universities. *Nature*, 479(7371), 15.

Calonico, S., Cattaneo, M. D., and Titiunik, R. (2014). Robust data-driven inference in the regression-discontinuity design. *Stata Journal*, 14(4), 909–946.

Camerer, C. F., Dreber, A., Forsell, E., Ho, T. H., Huber, J., Johannesson, M., . . . and Wu, H. (2016). Evaluating replicability of laboratory experiments in economics. *Science*, 351(6280), 1433–1436.

Camerer, C. F., Dreber, A., Holzmeister, F., Ho, T. H., Huber, J., Johannesson, M., . . . and Wu, H. (2018). Evaluating the replicability of social science experiments in *Nature* and *Science* between 2010 and 2015. *Nature Human Behaviour*, 2(9), 637–644.

Carlisle, J. B. (2012). The analysis of 168 randomised controlled trials to test data integrity. *Anaesthesia*, 67(5), 521–537.

Carsey, T. M. (2014). Making DA-RT a reality. *PS: Political Science and Politics*, 47(1), 72–77.

Cattaneo, M. D., and Titiunik, R. (2022). Regression discontinuity designs. *Annual Review of Economics*, 14, 821–851.

Cattaneo, M. D., and Vazquez-Bare, G. (2017). The choice of neighborhood in regression discontinuity designs. *Observational Studies*, 3(2), 134–146.

Caughey, D., and Sekhon, J. S. (2011). Elections and the regression discontinuity design: Lessons from close US house races, 1942–2008. *Political Analysis*, 19(4), 385–408.

Chambers, C. (2017). *The Seven Deadly Sins of Psychology: A Manifesto for Reforming the Culture of Scientific Practice*. Princeton University Press.

Chang, A. C., and Li, P. (2017). A preanalysis plan to replicate sixty economics research papers that worked half of the time. *American Economic Review*, 107(5), 60–64.

Chatfield, C. (1995). Model uncertainty, data mining and statistical inference. *Journal of the Royal Statistical Society Series A: Statistics in Society*, 158(3), 419–444.

Chawla, D. S. (2021). 8% of researchers in Dutch survey have falsified or fabricated data. *Nature: Career News*, July 22.

Chevassus-au-Louis, N. (2019). *Fraud in the Lab: The High Stakes of Scientific Research*. Harvard University Press.

Cho, W. K. T., and Gaines, B. J. (2007). Breaking the (Benford) law: Statistical fraud detection in campaign finance. *American Statistician*, 218–223.

Chopra, V., and Eagle, K. A. (2012). Perioperative mischief: The price of academic misconduct. *American Journal of Medicine*, 125(10), 953–955.

Christensen, B., and Christensen, S. (2014). Are female hurricanes really deadlier than male hurricanes? *Proceedings of the National Academy of Sciences*, 111(34), E3497–E3498.

Christensen, G., and Miguel, E. (2018). Transparency, reproducibility, and the credibility of economics research. *Journal of Economic Literature*, 56(3), 920–980.

Christensen, G., Freese, J., and Miguel, E. (2019). *Transparent and Reproducible Social Science Research: How to Do Open Science*. University of California Press.

Clark, W. R., and Golder, M. (2023). *Interaction Models*. Cambridge University Press.

Collier, R. (2015). Scientific misconduct or criminal offence? *Canadian Medical Association Journal* 17: 1273–1274.

Couzin-Frankel, J. (2013). The power of negative thinking. *Science* 342(6154): 68–69.

Crocker, J. (2011). The road to fraud starts with a single step. *Nature*, 479(7372), 151–151.

Dal-Ré, R., Bouter, L. M., Cuijpers, P., Gluud, C., and Holm, S. (2020). Should research misconduct be criminalized? *Research Ethics*, 174701611989840.

Dalton, R. (2005). Obesity expert owns up to million-dollar crime. *Nature*, 434(7032), 424–425.

Deckert, J., Myagkov, M., and Ordeshook, P. C. (2011). Benford's law and the detection of election fraud. *Political Analysis*, 19(3), 245–268.

Demir, B., and Javorcik, B. (2020). Trade policy changes, tax evasion and Benford's law. *Journal of Development Economics*, 144, 102456.

Diekmann, A. (2007). Not the first digit! Using Benford's law to detect fraudulent scientific data. *Journal of Applied Statistics*, 34(3), 321–329.

Donohue, J. J., III, and Levitt, S. D. (2001). The impact of legalized abortion on crime. *Quarterly Journal of Economics*, 116(2), 379–420.

Donohue, J. J., III, and Levitt, S. D. (2004). Further evidence that legalized abortion lowered crime. *Journal of Human Resources*, 39(1).

Donohue, J. J., and Wolfers, J. (2006a). Uses and abuses of empirical evidence in the death penalty debate. NBER working paper 11982.

Donohue, J. J., and Wolfers, J. (2006b). The death penalty: No evidence for deterrence. *Economists' Voice*, 3(5).

Durtschi, C., Hillison, W., and Pacini, C. (2004). The effective use of Benford's law to assist in detecting fraud in accounting data. *Journal of Forensic Accounting*, 5(1), 17–34.

Dyer, C. (2015). Migraine doctor is suspended for serious breach of professional standards. *BMJ* 2015; 350 (published February 20, 2015).

Eckel, C. C., and Grossman, P. J. (2008). Men, women and risk aversion: Experimental evidence. *Handbook of Experimental Economics Results* 1, 1061–1073.

Elliott, C. (2024). *The Occasional Human Sacrifice: Medical Experimentation and the Price of Saying No*. W. W. Norton.

Else, H. (2019). What universities can learn from one of science's biggest frauds. *Nature*, 570(7761), 287–289.

Else, H., and Van Noorden, R. (2021). The fight against fake-paper factories that churn out sham science. *Nature*, 591(7851), 516–520.

Enserink, M. (2012). Fraud-detection tool could shake up psychology. *Nature* 487, 18–19.

Esarey, J. (2017). Lowering the threshold of statistical significance to $p<0.005$ to encourage enriched theories of politics. *Political Methodologist*, 24(2), 13–19.

Esarey, J., and Danneman, N. (2015). A quantitative method for substantive robustness assessment. *Political Science Research and Methods*, 3(1), 95–111.

Fabbri, A., Lai, A., Grundy, Q., and Bero, L. A. (2018). The influence of industry sponsorship on the research agenda: A scoping review. *American Journal of Public Health*, 108(11), e9–e16.

Falk, A., and Heckman, J. J. (2009). Lab experiments are a major source of knowledge in the social sciences. *Science*, 326(5952), 535–538.

Fanelli, D. (2009). How many scientists fabricate and falsify research? A systematic review and meta-analysis of survey data. *PLOS One*, 4(5), e5738.

Feine, J., and Jakubovics, N. (2021). Science in the spotlight: A crisis of confidence? *JDR Clinical and Translational Research*, 6(1), 4–7.

Fewster, R. M. (2009). A simple explanation of Benford's law. *American Statistician*, 26–32.

Foote, C. L., and Goetz, C. F. (2005), Testing economic hypotheses with state level data: A comment on Donohue and Levitt (2001). Federal Reserve Bank of Boston Working Paper No. 05-15.

Francis, G. (2012). The psychology of replication and replication in psychology. *Perspectives on Psychological Science*, 7(6), 585–594.

Fraser, H., Barnett, A., Parker, T. H., and Fidler, F. (2020). The role of replication studies in ecology. *Ecology and Evolution*, 10(12), 5197–5207.

Fréchette, G. R. (2011). *Laboratory Experiments: Professionals Versus Students*. Available at SSRN 1939219.

Freedman, D. A. (1991). Statistical models and shoe leather. *Sociological Methodology*, 291–313.

Galbraith, K. L. (2017). Life after research misconduct: Punishments and the pursuit of second chances. *Journal of Empirical Research on Human Research Ethics*, 12(1), 26–32.

Gall, T., and Maniadis, Z. (2019). Evaluating solutions to the problem of false positives. *Research Policy*, 48(2), 506–515.

Gelman, A., and Carlin, J. (2014). Beyond power calculations: Assessing type S (sign) and type M (magnitude) errors. *Perspectives on Psychological Science*, 9(6), 641–651.

Gelman, A., Hill, J., and Vehtari, A. (2021). *Regression and Other Stories*. Cambridge University Press.

Gelman, A., and Imbens, G. (2019). Why high-order polynomials should not be used in regression discontinuity designs. *Journal of Business and Economic Statistics*, 37(3), 447–456.

Gelman, A., and Stern, H. (2006). The difference between "significant" and "not significant" is not itself statistically significant. *American Statistician*, 60(4), 328–331.

Gelman, A., and Loken, E. (2014). The statistical crisis in science. *American Scientist*, 102(6), 460–465.

George, S. L., and Buyse, M. (2015). Data fraud in clinical trials. *Clinical Investigation*, 5(2), 161.

Gerber, A. S., and Malhotra, N. (2008). Publication bias in empirical sociological research: Do arbitrary significance levels distort published results? *Sociological Methods and Research*, 37(1), 3–30.

Gill, J. (1999). The insignificance of null hypothesis significance testing. *Political Research Quarterly*, 52(3), 647–674.

Goodstein, D. (2010). *On Fact and Fraud*. Princeton University Press.

Gopalakrishna, G., Ter Riet, G., Vink, G., Stoop, I., Wicherts, J. M., and Bouter, L. M. (2022). Prevalence of questionable research practices, research misconduct and their potential explanatory factors: A survey among academic researchers in the Netherlands. *PLOS One*, 17(2), e0263023.

Greiner, T., and Pokorny, A. D. (1990). Can death be postponed? The death-dip phenomenon in psychiatric patients. *OMEGA—Journal of Death and Dying*, 20(2), 117–126.

Grimes, D. R., Bauch, C. T., and Ioannidis, J. P. (2018). Modelling science trustworthiness under publish or perish pressure. *Royal Society Open Science*, 5(1), 171511.

Gross, C. (2016). Scientific misconduct. *Annual Review of Psychology*, 67, 693–711.

Gross, J. H. (2015). Testing what matters (if you must test at all): A context-driven approach to substantive and statistical significance. *American Journal of Political Science*, 59(3), 775-788.

Harris, R. (2017). *Rigor Mortis: How Sloppy Science Creates Worthless Cures, Crushes Hope, and Wastes Billions*. Hachette UK.

Hartgerink, C. H., Van Aert, R. C., Nuijten, M. B., Wicherts, J. M., and Van Assen, M. A. (2016). Distributions of p-values smaller than. 05 in psychology: What is going on? *PeerJ*, 4, e1935.

Hauser, M. D., Yang, C., Berwick, R. C., Tattersall, I., Ryan, M. J., Watumull, J., . . . and Lewontin, R. C. (2014a). The mystery of language evolution. *Frontiers in Psychology*, 401.

Head, M. L., Holman, L., Lanfear, R., Kahn, A. T., and Jennions, M. D. (2015). The extent and consequences of p-hacking in science. *PLOS Biology*, 13(3), e1002106.

Hearn, P. D. (2009). *Hurricane Camille: Monster Storm of the Gulf Coast*. University Press of Mississippi.

Heckman, J. J., and Urzua, S. (2010). Comparing IV with structural models: What simple IV can and cannot identify. *Journal of Econometrics*, 156(1), 27–37.

Heckman, J. J., Urzua, S., and Vytlacil, E. (2006). Understanding instrumental variables in models with essential heterogeneity. *Review of Economics and Statistics*, 88(3), 389–432.

Herndon, T., Ash, M., and Pollin, R. (2014). Does high public debt consistently stifle economic growth? A critique of Reinhart and Rogoff. *Cambridge Journal of Economics*, 38(2), 257–279.

Ho, D. E., Imai, K., King, G., and Stuart, E. A. (2007). Matching as nonparametric preprocessing for reducing model dependence in parametric causal inference. *Political Analysis*, 15(3), 199–236.

Holden, C. (1987). Breuning's accuser has grant deferred. *Science*, 236(4798), 144.

Holden, K. L. (2016). Buy the book? Evidence on the effect of textbook funding on school-level achievement. *American Economic Journal: Applied Economics*, 100–127.

Holland, P. W. (1986). Statistics and causal inference. *Journal of the American Statistical Association*, 945–960.

Holm, S., and Hofmann, B. (2018). Associations between attitudes towards scientific misconduct and self-reported behavior. *Accountability in Research*, 25(5), 290–300.

Honeycutt, N., and Jussim, L. (2020). A model of political bias in social science research. *Psychological Inquiry*, 31(1), 73–85.

Horton, R. (2016). Offline: Paolo Macchiarini: Science in conflict. *The Lancet*, 387(10020), 732.

Imai, K., Tingley, D., and Yamamoto, T. (2013). Experimental designs for identifying causal mechanisms. *Journal of the Royal Statistical Society: Series A (Statistics in Society)*, 176(1), 5–51.

Imbens, G. W. (2010). Better LATE than nothing: Some comments on Deaton (2009) and Heckman and Urzua (2009). *Journal of Economic Literature*, 48(2), 399–423.

Imbens, G. W. (2021). Statistical significance, p-values, and the reporting of uncertainty. *Journal of Economic Perspectives*, 35(3), 157–174.

Imbens, G. W., and Rubin, D. B. (2015). *Causal Inference in Statistics, Social, and Biomedical Sciences*. Cambridge University Press.

Ioannidis, J. P. (2005). Why most published research findings are false. *PLOS Medicine*, 2(8), e124.

Ioannidis, J. P. (2012). Why science is not necessarily self-correcting. *Perspectives on Psychological Science*, 7(6), 645–654.

Jackson, C. K. (2020). Does school spending matter? The new literature on an old question. NBER working paper 25368.

Jamieson, K. H. (2018). Crisis or self-correction: Rethinking media narratives about the well-being of science. *Proceedings of the National Academy of Sciences*, 115(11), 2620–2627.

Jones, D. N., and Paulhus, D. L. (2014). Introducing the short dark triad (SD3) a brief measure of dark personality traits. *Assessment*, 21(1), 28–41.

Jung, K., Shavitt, S., Viswanathan, M., and Hilbe, J. M. (2014b). Female hurricanes are deadlier than male hurricanes. *Proceedings of the National Academy of Sciences*, 111(24), 8782–8787.

Kaiser, J. (2023). "I should have done better." Stanford head steps down. *Science*, 381(6656), 366–367.

Kalichman, M. W., and Friedman, P. J. (1992). A pilot study of biomedical trainees' perceptions concerning research ethics. *Academic Medicine*, 67(11), 769–775.

Kaplan, R. M., and Irvin, V. L. (2015). Likelihood of null effects of large NHLBI clinical trials has increased over time. *PLOS One*, 10(8), e0132382.

Kelly, C. D. (2019). Rate and success of study replication in ecology and evolution. *PeerJ*, 7, e7654.

Kenny, K., Fotaki, M., and Scriver, S. (2019). Mental health as a weapon: Whistleblower retaliation and normative violence. *Journal of Business Ethics*, 160, 801–815.

Kerr, N. L. (1998). HARKing: Hypothesizing after the results are known. *Personality and Social Psychology Review*, 2(3), 196–217.

King, G., and Nielsen, R. (2019). Why propensity scores should not be used for matching. *Political Analysis*, 27(4), 435–454.

King, G., and Zeng, L. (2007). Detecting model dependence in statistical inference: A response. *International Studies Quarterly*, 51(1), 231–241.

Klein, R. A., Ratliff, K. A., Vianello, M., Adams Jr, R. B., Bahník, Š., Bernstein, M. J., . . . and Nosek, B. A. (2014). Investigating variation in replicability. *Social Psychology* 45:3, 137–252.

Kranke, P., Apfel, C. C., and Roewer, N. (2000). Reported data on granisetron and postoperative nausea and vomiting by Fujii et al. are incredibly nice! *Anesthesia and Analgesia*, 90(4), 1004–1006.

Kristal, A. S., Whillans, A. V., Bazerman, M. H., Gino, F., Shu, L. L., Mazar, N., and Ariely, D. (2020). Signing at the beginning versus at the end does not decrease dishonesty. *Proceedings of the National Academy of Sciences*, 117(13), 7103–7107.

Krolzig, H. M., and Hendry, D. F. (2001). Computer automation of general-to-specific model selection procedures. *Journal of Economic Dynamics and Control*, 25(6–7), 831–866.

Kuhar, M. J. (2009). Blacklisting among scientists. *Synapse*, 63(7), 539.

Kupferschmidt, K. (2018). Researcher at the center of an epic fraud remains an enigma to those who exposed him. *Science*, 80, 636–641.

Laitin, D. D., and Reich, R. (2017). Trust, transparency, and replication in political science. *PS: Political Science and Politics*, 50(1), 172–175.

Larcker, D. F., and Rusticus, T. O. (2010). On the use of instrumental variables in accounting research. *Journal of Accounting and Economics*, 49(3), 186–205.

Leamer, E. E. (1978). *Specification Searches: Ad Hoc Inference with Nonexperimental Data*. Wiley.

Leamer, E. E. (1985). Sensitivity analyses would help. *American Economic Review*, 75(3), 308–313.

Lee, D. S., and Lemieux, T. (2010). Regression discontinuity designs in economics. *Journal of Economic Literature*, 48(2), 281–355.

Lee, S. M. (2018). Turns out Cornell's Brian Wansink investigation is still very much going. buzzfeednews.com/article/stephaniemlee/brian-wansink-cornell-investigation-continues.

Lee, S. M. (2023). The Harvard fraud expert accused of committing fraud: Francesca Gino had a hot streak of buzzy research that made her a star. Was it a warning sign? *Chronicle of Higher Education*, 69(22), 16–23.

Lesné, S., Koh, M. T., Kotilinek, L., Kayed, R., Glabe, C. G., Yang, A., Gallagher, M., and Ashe, K. H. (2024). Retraction note: A specific amyloid-β protein assembly in the brain impairs memory. *Nature*, 631(240)

Levitt, S. D., and List, J. A. (2007). What do laboratory experiments measuring social preferences reveal about the real world? *Journal of Economic Perspectives*, 21(2), 153–174.

Lock, S., and Lock, S. (1997). Fraud in medical research. *NT Research*, 2(3), 161–163.

Lubalin, J. S., and Matheson, J. L. (1999). The fallout: What happens to whistleblowers and those accused but exonerated of scientific misconduct? *Science and Engineering Ethics*, 5, 229–250.

Malek, J. (2010). To tell or not to tell? The ethical dilemma of the would-be whistleblower. *Accountability in Research*, 17(3), 115–129.

Maley, S. (2014). Statistics show no evidence of gender bias in the public's hurricane preparedness. *Proceedings of the National Academy of Sciences*, 111(37), E3834.

Malter, D. (2014). Female hurricanes are not deadlier than male hurricanes. *Proceedings of the National Academy of Sciences*, 111(34), E3496.

Manski, C. F. (2013). Identification of treatment response with social interactions. *Econometrics Journal*, 16(1), S1–S23.

Manzi, J. (2012). *Uncontrolled: The Surprising Payoff of Trial-and-Error for Business, Politics, and Society*. Basic Books.

Markowitz, D. M., and Hancock, J. T. (2014). Linguistic traces of a scientific fraud: The case of Diederik Stapel. *PLOS One*, 9(8), e105937.

Markowitz, D. M., and Hancock, J. T. (2016). Linguistic obfuscation in fraudulent science. *Journal of Language and Social Psychology*, 35(4), 435–445.

Martin, B. (1992). Scientific fraud and the power structure of science. *Prometheus*, 10(1), 83–98.

Masicampo, E. J., and Lalande, D. R. (2012). A peculiar prevalence of p values just below .05. *Quarterly Journal of Experimental Psychology*, 65(11), 2271–2279.

Mastroianni, A. (2022). The rise and fall of peer review: Why the greatest scientific experiment in history failed, and why that's a great thing. *Experimental History*, December 13. experimental-history.com/p/the-rise-and-fall-of-peer-review.

McAleer, M. (1994). Sherlock Holmes and the search for truth: A diagnostic tale. *Journal of Economic Surveys*, 8(4), 317–370.

McIntyre, L. (2019). *The Scientific Attitude: Defending Science from Denial, Fraud, and Pseudoscience*. MIT Press.

McNutt, M. (2020). *Self-Correction by Design*. MIT Press.

McShane, B. B., Gal, D., Gelman, A., Robert, C., and Tackett, J. L. (2019). Abandon statistical significance. *American Statistician*, 73(suppl. 1), 235–245.

Mellon, J. (2025). Rain, rain, go away: 194 potential exclusion-restriction violations for studies using weather as an instrumental variable. *American Journal of Political Science*, 69(3), 881–898.

Merton, R. K. (1957). Priorities in scientific discovery: A chapter in the sociology of science. *American Sociological Review*, 22(6), 635–665.

Miranda-Zanetti, M., Delbianco, F., and Tohmé, F. (2019). Tampering with inflation data: A Benford law–based analysis of national statistics in Argentina. *Physica A: Statistical Mechanics and Its Applications*, 525, 761–770.

Munafò, M. R., Nosek, B. A., Bishop, D. V., Button, K. S., Chambers, C. D., Percie du Sert, N., . . . and Ioannidis, J. (2017). A manifesto for reproducible science. *Nature Human Behaviour*, 1(1), 1–9.

Nature Neuroscience (2006). Can peer review police fraud? (Editorial). *Nature Neuroscience* 9(149).

Nesterak, E. (2023). Amid uncertainty about Francesca Gino's research, the many co-authors project could provide clarity. *Behavioral Scientist*, August 30. behavioralscientist.org/amid-uncertainty-about-francesca-gino-led-research-the-many-co-authors-project-could-provide-clarity/.

Neumayer, E. (2005). Inequality and violent crime: Evidence from data on robbery and violent theft. *Journal of Peace Research*, 42(1), 101–112.

Neumayer, E., and Plümper, T. (2017). *Robustness Tests for Quantitative Research*. Cambridge University Press.

Newhouse, J. P., and Insurance Experiment Group Staff. (1993). *Free for All? Lessons from the RAND Health Insurance Experiment*. Harvard University Press.

Nosek, B. A., and Lakens, D. (2014). Registered reports. *Social Psychology* 45(3).

Nosek, B. A., Ebersole, C. R., DeHaven, A. C., and Mellor, D. T. (2018). The preregistration revolution. *Proceedings of the National Academy of Sciences*, 115(11), 2600–2606.

Nosek, B. A., Hardwicke, T. E., Moshontz, H., Allard, A., Corker, K. S., Dreber, A., . . . and Vazire, S. (2022). Replicability, robustness, and reproducibility in psychological science. *Annual Review of Psychology*, 73, 719–748.

Nunn, Clyde Z. (1977). Is there a crisis of confidence in science? *Science* 198(4321): 995.

Nyman, J. A. (2007). American health policy: Cracks in the foundation. *Journal of Health Politics, Policy and Law*, 32(5), 759–783.

OECD (2007). *Best Practices for Ensuring Scientific Integrity and Preventing Misconduct*. OECD.

Open Science Collaboration. (2015). Estimating the reproducibility of psychological science. *Science* 349(6251), aac4716.

Panhans, M. T., and Singleton, J. D. (2017). The empirical economist's toolkit: From models to methods. *History of Political Economy*, 49(suppl.), 127–157.

Papageorgiou, S. N., Xavier, G. M., Cobourne, M. T., and Eliades, T. (2018). Registered trials report less beneficial treatment effects than unregistered ones: A meta-epidemiological study in orthodontics. *Journal of Clinical Epidemiology*, 100, 44–52.

Parker, L., Boughton, S., Lawrence, R., and Bero, L. (2022). Experts identified warning signs of fraudulent research: A qualitative study to inform a screening tool. *Journal of Clinical Epidemiology*, 151, 1–17.

Penders, B., and Shaw, D. M. (2020). Civil disobedience in scientific authorship: Resistance and insubordination in science. *Accountability in Research*, 27(6), 347–371.

Pham, M. T., and Oh, T. T. (2021). On not confusing the tree of trustworthy statistics with the greater forest of good science: A comment on Simmons et al.'s perspective on pre-registration. *Journal of Consumer Psychology*, 31(1), 181–185.

Phillips, D., and King, E. (1988). Death takes a holiday: Mortality surrounding major social occasions. *The Lancet*, 332(8613), 728–732.

Phillips, D. P., and Feldman, K. A. (1973). A dip in deaths before ceremonial occasions: Some new relationships between social integration and mortality. *American Sociological Review*, 38(6), 678–696.

Phillips, D. P., and Smith, D. G. (1990). Postponement of death until symbolically meaningful occasions. *JAMA*, 263(14), 1947–1951.

Piketty, T. (2014). *Capital in the Twenty-First Century*. Harvard University Press.

Plümper, T., Flores, A. Q., and Neumayer, E. (2017). The double-edged sword of learning from disasters: Mortality in the Tohoku tsunami. *Global Environmental Change*, 44, 49–56.

Prinz, F., Schlange, T., and Asadullah, K. (2011). Believe it or not: How much can we rely on published data on potential drug targets? *Nature Reviews Drug Discovery*, 10(9), 712.

Pritchett, L., and Sandefur, J. (2015). Learning from experiments when context matters. *American Economic Review*, 105(5), 471–475.

Rasool, S. I., and Schneider, S. H. (1971). Atmospheric carbon dioxide and aerosols: Effects of large increases on global climate. *Science*, 173(3992), 138–141.

Reinhart, C. M., and Rogoff, K. S. (2010). Growth in a time of debt. *American Economic Review*, 100(2), 573–578.

Rhodes, R., and Strain, J. J. (2004). Whistleblowing in academic medicine. *Journal of Medical Ethics*, 30(1), 35–39.

Ritchie, S. (2021). *Science Fictions: Exposing Fraud, Bias, Negligence and Hype in Science*. Vintage.

Rosenbaum, P. (2002). *Observational Studies*. Springer.

Rosenbaum, P. (2010). *Design of Observational Studies*. Springer.

Rosenbaum, P. R., and Rubin, D. B. (1983). The central role of the propensity score in observational studies for causal effects. *Biometrika*, 70(1), 41–55.

Roth, W., Tam, M., Bi, C., Kim, J., Lewis, J., Ho, R., and Apollonio, D. E. (2022). Changes in California cannabis exposures following recreational legalization and the COVID-19 pandemic. *Clinical Toxicology*, 60(5), 632–638.

Rothschild, J. (2013). The fate of whistleblowers in nonprofit organizations. *Nonprofit and Voluntary Sector Quarterly*, 42(5), 886–901.

Rouder, J. N., and Morey, R. D. (2011). A Bayes factor meta-analysis of Bem's ESP claim. *Psychonomic Bulletin and Review*, 18, 682–689.

Royer, H., and Smith, G. (1998). Can the famous really postpone death? *Social Biology*, 45(3–4), 302–305.

Rubin, D. B. (1974). Estimating causal effects of treatments in randomized and nonrandomized studies. *Journal of Educational Psychology*, 66(5), 688.

Rubin, D. B. (2008). Comment: The design and analysis of gold standard randomized experiments. *Journal of the American Statistical Association*, 103(484), 1350–1356.

Sackrowitz, H., and Samuel-Cahn, E. (1999). P values as random variables-expected p values. *American Statistician*, 326–331.

Salemink, O., and Verrips, J. (2020). More thoughts on scholarly misconduct, and what we can learn from it: A critical response to Peter Jan Margry's article about the Bax affair. *Ethnologia Europaea*, 49(2), 145–155.

Samii, C. (2016). Causal empiricism in quantitative research. *Journal of Politics*, 78(3), 941–955.

Scheel, A. M., Schijen, M. R., and Lakens, D. (2021). An excess of positive results: Comparing the standard psychology literature with registered reports. *Advances in Methods and Practices in Psychological Science*, 4(2), 25152459211007467.

Schimmack, U. (2012). The ironic effect of significant results on the credibility of multiple-study articles. *Psychological Methods*, 17(4), 551.

Sekhon, J. S. (2009). Opiates for the matches: Matching methods for causal inference. *Annual Review of Political Science*, 12(1), 487–508.

Sekhon, J. S., and Titiunik, R. (2017). On interpreting the regression discontinuity design as a local experiment. In Cattaneo, M. D., and Escanciano, J. C. (eds.), *Regression Discontinuity Designs*. Emerald Publishing.

Shamoo, A. E., and Resnik, D. B. (2015). *Responsible Conduct of Research*. 3rd ed. Oxford University Press.

Silva, L., and Figueiredo Filho, D. (2021). Using Benford's law to assess the quality of COVID-19 register data in Brazil. *Journal of Public Health*, 43(1), 107–110.

Simmons, J. P., Nelson, L. D., and Simonsohn, U. (2011). False-positive psychology: Undisclosed flexibility in data collection and analysis allows presenting anything as significant. *Psychological Science*, 22(11), 1359–1366.

Simmons, J. P., Nelson, L. D., and Simonsohn, U. (2012). A 21 word solution. SSRN 2160588.

Simmons, J. P., Nelson, L. D., and Simonsohn, U. (2021a). Pre-registration: Why and how. *Journal of Consumer Psychology*, 31(1), 151–162.

Simmons, J. P., Nelson, L. D., and Simonsohn, U. (2021b). Pre-registration is a game changer. But, like random assignment, it is neither necessary nor sufficient for credible science. *Journal of Consumer Psychology*, 31(1), 177–180.

Simonsohn, U. (2013). Just post it: The lesson from two cases of fabricated data detected by statistics alone. *Psychological Science*, 24(10), 1875–1888.

Simonsohn, U. (2014). Posterior-hacking: Selective reporting invalidates Bayesian results also. Available at dx.doi.org/10.2139/ssrn.2374040.

Simonsohn, U., Nelson, L. D., and Simmons, J. P. (2014). P-curve: a key to the file-drawer. *Journal of Experimental Psychology: General*, 143(2), 534.

Simonsohn, U., Simmons, J. P., and Nelson, L. D. (2020). Specification curve analysis. *Nature Human Behaviour*, 4(11), 1208–1214.

Smaldino, P. E., and McElreath, R. (2016). The natural selection of bad science. *Royal Society Open Science*, 3(9), 160384.

Smith, G. (2014). *Standard Deviations: Flawed Assumptions, Tortured Data and Other Ways to Lie with Statistics*. Duckworth.

Smith, G. (2016). Hurricane names: A bunch of hot air? *Weather and Climate Extremes*, 12, 80–84.

Stapel, D. (2014). Faking science: A true story of academic fraud. Trans. N. J. L. Brown. Unpublished manuscript. nick.brown.free.fr/stapel/FakingScience-20161115 .pdf.

Steegen, S., Tuerlinckx, F., Gelman, A., and Vanpaemel, W. (2016). Increasing transparency through a multiverse analysis. *Perspectives on Psychological Science*, 11(5), 702–712.

Steen, R. G. (2011). Retractions in the scientific literature: Is the incidence of research fraud increasing? *Journal of Medical Ethics*, 37(4), 249–253.

Steen, R. G., Casadevall, A., and Fang, F. C. (2016). Why has the number of scientific retractions increased? *PLOS One* 8(7), e68397.

Stommes, D., Aronow, P. M., and Sävje, F. (2021). On the reliability of published findings using the regression discontinuity design in political science. arXiv preprint. arXiv:2109.14526.

Swarecki, B. (2014). "Female" hurricanes are deadlier. *Science,* June 2, https://www.science.org/content/article/female-hurricanes-are-deadlier.

Tavare, A. (2011). Managing research misconduct: Is anyone getting it right? *BMJ,* 343.

Tijdink, J. K., Bouter, L. M., Veldkamp, C. L., van de Ven, P. M., Wicherts, J. M., and Smulders, Y. M. (2016). Personality traits are associated with research misbehavior in Dutch scientists: A cross-sectional study. *PLOS One* 11(9), e0163251.

Ting, M. M. (2008). Whistleblowing. *American Political Science Review,* 102(2), 249–267.

University of Würzburg. (2012). Press release: Too good to be true. uni-wuerzburg.de/en/news-and-events/news/detail/news/too-good-to-be-true/. URL no longer extant.

Urquiola, M., and Verhoogen, E. (2009). Class-size caps, sorting, and the regression-discontinuity design. *American Economic Review,* 99(1), 179–215.

Van der Zee, T. (2017). The Wansink dossier: An overview. *The Skeptical Scientist.* timvanderzee.com/the-wansink-dossier-an-overview/.

Van Noorden, R. (2015). Retracted gay-marriage study debated at misconduct meet-up. *Nature,* 522(7544), 14–16.

Varian, H. R. (1972). Benford's law. *American Statistician,* 26(3), 65.

Vazire, S. (2019). Do we want to be credible or incredible? *APS Observer,* 33.

Vie, K. J. (2020). How should researchers cope with the ethical demands of discovering research misconduct? Going beyond reporting and whistle-blowing. *Life Sciences, Society and Policy,* 16(1), 1–18.

Vogel, G. (2011). Psychologist accused of fraud on "astonishing scale." *Science* 334(6056), 579.

Vogel, G. (2013). Trachea transplants test the limits. *Science,* 340(6130): 266–268.

von Hippel, P. T. (2022). Is psychological science self-correcting? Citations before and after successful and failed replications. *Perspectives on Psychological Science,* 17(6), 1556–1565.

Wagenmakers, E. J., Wetzels, R., Borsboom, D., and Van Der Maas, H. L. (2011). Why psychologists must change the way they analyze their data: The case of psi: Comment on Bem (2011). *Journal of Personality and Social Psychology,* 100(3), 426–432.

Wager, E. (2013). The UK should lead the way on research integrity. *BMJ,* 346, f2348.

Waltz, E. (2022). Elizabeth Holmes verdict: Researchers share lessons for science. *Nature*, 601(7892), 173–174.

Wasserstein, R. L., and Lazar, N. A. (2016). The ASA statement on p-values: Context, process, and purpose. *American Statistician*, 70(2), 129–133.

Webb, G. P. (2021). *Error and Fraud: The Dark Side of Biomedical Research*. CRC Press.

Weed, D. L. (1998). Preventing scientific misconduct. *American Journal of Public Health*, 88(1), 125–129.

White, C. (2015). Ranjit Chandra: How reputation bamboozled the scientific community. *BMJ*, 351.

Winking, J., and Mizer, N. (2013). Natural-field dictator game shows no altruistic giving. *Evolution and Human Behavior*, 34(4), 288–293.

Winship, C., and Morgan, S. L. (1999). The estimation of causal effects from observational data. *Annual Review of Sociology*, 25(1), 659–706.

Young, A. (2022). Consistency without inference: Instrumental variables in practical application. *European Economic Review*, 147, 104112.

Young, D. C., and Hade, E. M. (2004). Holidays, birthdays, and postponement of cancer death. *JAMA*, 292(24), 3012–3016.

Young, J. R. (2008). Journals find many images in research are faked. *Chronicle of Higher Education*, 54(39).

Ziliak, S. T., and McCloskey, D. N. (2013). We agree that statistical significance proves essentially nothing: A rejoinder to Thomas Mayer. *Econ Journal Watch*, 10(1), 97.

Retracted Articles Cited

Gino, F., Kouchaki, M., and Casciaro, T. (2020). Why connect? Moral consequences of networking with a promotion or prevention focus. *Journal of Personality and Social Psychology*, 119(6), 1221.

Gino, F., and Wiltermuth, S. S. (2014). Evil genius? How dishonesty can lead to greater creativity. *Psychological Science*, 25(4), 973–981.

LaCour, M. J., and Green, D. P. (2014). When contact changes minds: An experiment on transmission of support for gay equality. *Science*, 346(6215), 1366–1369.

Lammers, J., Stapel, D. A., and Galinsky, A. D. (2010). Power increases hypocrisy: Moralizing in reasoning, immorality in behavior. *Psychological Science*, 21(5), 737–744.

Marx, D. M., Stapel, D. A., and Muller, D. (2005). We can do it: The interplay of construal orientation and social comparisons under threat. *Journal of Personality and Social Psychology*, 88(3), 432.

Shu, L. L., Mazar, N., Gino, F., Ariely, D., and Bazerman, M. H. (2012). Signing at the beginning makes ethics salient and decreases dishonest self-reports in comparison to signing at the end. *Proceedings of the National Academy of Sciences*, 109(38), 15197–15200.

Smeesters, D., and Liu, J. (2011). The effect of color (red versus blue) on assimilation versus contrast in prime-to-behavior effects. *Journal of Experimental Social Psychology*, 47(3), 653–656.

Stapel, D. A., and Koomen, W. (2001). I, we, and the effects of others on me: How self-construal level moderates social comparison effects. *Journal of Personality and Social Psychology*, 80(5), 766.

Wakefield, A. J., Murch, S. H., Anthony, A., Linnell, J., Casson, D. M., Malik, M., . . . and Walker-Smith, J. A. (1998). Ileal-lymphoid-nodular hyperplasia, non-specific colitis, and pervasive developmental disorder in children. *The Lancet*, 351(9103), 637–641.

Index

Note: Italicized page locators refer to figures; tables are noted with a *t*.

Publisher contact:
The MIT Press
Massachusetts Institute of Technology
77 Massachusetts Avenue, Cambridge, MA 02139
mitpress.mit.edu

EU Authorised Representative:
Easy Access System Europe, Mustamäe tee 50,
10621 Tallinn, Estonia
gpsr.requests@easproject.com

Printed by Integrated Books International,
United States of America